How to Farm for Profit

How to Farm for Profit

Practical Enterprise Analysis

Donald M. Fedie
with the collaboration of Michael H. Prosser, CPA

Iowa State University Press • Ames

Donald M. Fedie, a financial and management advisor, is president of Agri Control. He has published numerous articles on agricultural management. This book is the outgrowth of his weekly column for the electronic issue of FarmDayta Network.

© 1997 Iowa State University Press, Ames, Iowa 50014

♾ Printed on acid-free paper in the United States of America

First edition, 1997

Library of Congress Cataloging-in-Publication Data

Fedie, Donald M.
    How to farm for profit: practical enterprise analysis / Donald M. Fedie: with the collaboration of Michael H. Prosser, CPA.—1st ed.
        p.   cm.
    Includes bibliographical references and index.
    ISBN 0-8138-2560-1 (alk. paper)
    1. Agriculture—Accounting. 2. Agriculture—Finance. 3. Farm produce—Marketing. 4. Agriculture—Economic aspects—Forecasting. I. Title.
HF5686.A36F43   1997
630′.68′1—dc21
                                                                97-1053

Last digit is the print number: 9   8   7   6   5   4   3   2   1

✐ Contents

✐ Preface

EXPERIENCE AND THE PASSAGE of time continue to prove that the principles of this book are timeless and essential to the proper understanding of how a business operates and that they work for all businesses regardless of the industry with which they are associated. My career in the world of financial and management advising began in the 1960s when I worked with industrial manufacturing and food processing clients. During that time I was a member of a team that focused much of its energy on merger/acquisition candidates and companies that were financially pressed. Regardless of whether the company was being considered for acquisition or trying desperately to get on top of its financial future, we would begin our work process by examining past operations for a four- to five-year period. This examination allowed us to understand how the company had operated in the past and in essence to evaluate the strong and weak points of its operations. From this understanding we would formulate an approach for the company's future. Within this approach we would consider which operating parameters might have to change in order to match the company with a potential marriage partner or refinance for the future.

During the early 1970s, when I chose to return to and specialize in the field of agriculture, I carried with me this formula approach to understanding how a business works and how one can improve future operations. In part because the financial world of agriculture has changed dramatically during the past 25 years, the basic principles of business analysis (as applied to agriculture) have become solidly recognized as valid, helpful, and in fact imperative to the financial success of agriculture. These principles are embodied in the following four major analytical and forecasting functions: (1) Financial and enterprise analysis: The analysis of past operations, in terms of both financial history and operating performance. (2) A business forecast: Use of the operating factors analyzed to perform sensitivity analysis and to properly and effectively forecast a plan of operations for the future. (3) A marketing plan: Use of the forecast and sensitivity analysis to formulate and substantiate a plan for marketing, including the use of marketing futures to aid in protecting profits. (4) An accounting system: Use of an accounting system to accurately document and demonstrate what is happening to the business during the course of an operating year and to facilitate year-end operational analysis.

An honest examination of these major functions would lead one to conclude their equal importance. However, many segments of agriculture have historically considered the last three functions (forecasting, a marketing plan, and an accounting system) to be more important than the first. Agricultural producers have for years been encouraged to complete a cash flow budget, a forecast of operations. More recently the producer has been inundated with marketing concepts and tools and the need to use the commodity futures within a designed marketing strategy. In the past 10 years literally hundreds of software accounting systems have been designed for the agricultural producer, regardless of the fact that few understand computers, software, or accounting principles or how to interpret the results. The producer has all this at his/her disposal—without any idea or appreciation of what has happened during past operations.

It is imperative that one understand past operations to accurately forecast future operations, design a strategic marketing plan, and use the commodity futures to protect profits. Granted, an accounting system will aid in the analysis of past operations, if in fact it is designed to accomplish this task and one has the luxury of several years to gather information. My regard for historical financial and enterprise analysis began during the late 1960s before I came into agriculture. This regard continued to grow during the 1970s and reached its peak during the early 1980s—the beginning of the farm crisis. My interest and concern were heightened by the fact that many producers were being asked to vacate (eliminate) entire enterprises for the sake of reducing expenses and borrowed money. Neither they nor those asking for the reduction had documentation about whether the enterprise in question was the cause of negative earnings, whether the reduction in expenses and borrowed money would actually result in net positive earnings, how operations of the enterprises were interrelated, or whether the problem might be another nonsuspect enterprise.

With this realization our work for clients took on an entirely new dimension. We discovered that having performed enterprise analysis for a client, we could change the minds of those asking for the elimination of certain enterprises. Indeed, we discovered that those asking for a reduction (usually lenders) responded intelligently to good business analysis procedures. They were more than willing to accept the conclusions and provide adequate funds for a well-thought-out and carefully forecasted plan of attack. The word spread from one lender to another, and before long we were programming the first generation of a set of software systems to aid in analyzing the operations of a continually increasing number of producers.

Since the early 1980s we have financially analyzed the enterprise operations of several hundred producers on behalf of them and their lenders. I take a great deal of pride in knowing that many farmers are still in business today, doing well and being profitable, because we intervened at an important time in their financial life. "Doing well and being profitable" is the most important part of the previous sentence. The producers we have worked with are doing well and being profitable not because of some budgeting or marketing magic potion we served up but because we insisted on subjecting their business operations to historical financial and enterprise analysis before determining how they should change their operations. Once we and the producer understood from analysis what the break-even cost of doing business was, we completed a well-thought-out and achievable forecast of operations and applied appropriate marketing protections within a designed marketing plan. Having completed a forecast of operations, we included it as the first annual entry within the accounting system so as to provide for a monthly budget-to-actual comparison.

Understanding profitability and helping to make American

ag producers profitable bring us full circle. Profitability is still the greatest problem in agriculture today. We have personally been able to influence many operators over the past 20 years, but this merely scratches the surface. We hope that in reading this book you will become interested in the enterprise analysis process and in so doing will become conscious of the importance of profitability. Please allow me to emphasize that it is not enough to simply understand financial ratios and indicators; only enterprise analysis will allow you to understand the cost of production processes and how the enterprises relate to each other.

Its not an exaggeration to say that the survival of production agriculture as we know it depends entirely on individual producers knowing their exact break-even cost of production on each and every product. I hope you enjoy reading this book and profit from it—by becoming more profitable in your own business.

This book is more than just a recent venture. It's the result of a lifetime of learning and the practice of management sciences. Many people have been important and have contributed to that learning process and this book. First, let me thank Mr. Raymond Flanagan for taking this young kid under his arm—and teaching me all about the business of financial and management advising. One cannot learn this business in a school of higher education or from a book. One must have a sponsor, a teacher, an advocate, a father figure, a supreme believer. Ray was all of these, as well as a Duke-educated structural engineer who performed more calculations more quickly on an old-fashioned slide rule than most people can with a calculator. I will be forever grateful for his understanding and patience, knowledge and experience, and the fact that without any pretense he chose to share them with me.

Also, I wish to thank Mike Prosser, CPA, for his assistance and collaboration in writing this book. My thanks to Burton Pflueger, PhD; Ron Orth, PhD; Robert Strachan; Robert and Van Fishback; Greg Stine; Jim Bodyfield; Jeff Davis; Steve Almos; Lawrence Paulsen; Dean Chapman; Mike Wegner; and too many others to mention, for their collaboration and support over the years.

A personal note of thanks to my mother for teaching me patience, and to my dad for teaching me the value of hard work—and who preached one statement often enough to make it stick: "If you're going to take the time to do something, take enough time to do it right the first time."

℘ Introduction

AS A RESULT OF THE FARM CRISIS of the mid-1980s, roughly one-third of our agricultural producers were eliminated from the marketplace. Many of these casualties were farmers whose operations had been unprofitable because of inefficient use of capital assets, unenlightened marketing strategies for the commodities they produced, or simply poor fiscal management (no knowledge of break-even points, no projection of cash flow, or insufficient knowledge of agronomy). Because of the banking industry's then-prevailing practice of lending money based on the perceived value of assets rather than on payback capacity out of profits, marginal producers were permitted to continue operating financed by increased borrowings collateralized by land whose values continued to experience double-digit annual increases. Once the value of collateral plummeted, the party was over.

Given the complexities of agriculture as an industry—with factors such as weather, erosion, groundwater quality, government programs, and volatile market prices—it is obvious that the surviving agricultural producers must treat their enterprises as a business requiring skilled management and a basic understanding of financing to remain competitive. I anticipate that agricultural producers of the future will be much larger and tend to specialize more than in the past, will be more cognizant of the costs of production, will seek to minimize market risks through the use of futures contracts or options, will be more concerned about the long-term environmental effects of their farming methods, and will be motivated to maximize profits.

Ag lenders will continue to deemphasize asset-based lending practices, instead concentrating on customer profitability, with the more enlightened lenders actually serving as management partners with their loan customers. A more standardized approach to financial reporting and ratio analysis is a necessary component of successful business management. A thorough understanding of past operations is the cornerstone for accurate forecasting of future operations.

This book begins with a discussion of standardized financial reporting, dealing primarily with what has occurred in the past. Among the vital pieces of information to be derived from this process are the extent of profitability (or lack thereof), the sources and uses of cash during the reporting period, and the changes in owner's equity for the period. The second chapter of the book deals with ratio analysis, which describes the process of analyzing the information presented in financial reports for the purpose of gauging liquidity, debt repayment capacity, and return on assets and identifying relationships between certain elements of the financial statements that may be out of line with expected results. Expected results would be based on the history of the producer and the results of other producers in similar enterprise structures. Although the agricultural sector has in the past found it difficult to develop a database from which such normal or expected ratios can be extracted, movement toward this end has begun.

Much of the discussion contained in the first two chapters of the book reflects the recommendations of the Farm Financial Standards Council in their January 1995 exposure draft document titled *Financial Guidelines for Agricultural Producers*. The council, formerly named the Farm Financial Standards Task Force, is composed of volunteers representing the academic community, commercial banks, insurance companies, the farm credit system, FmHA, the accounting profession and financial advisors, and regulatory agencies. The exposure draft of *Financial Guidelines for Agricultural Producers* has three main sections that include recommendations on (1) universal financial reports, (2) universal financial criteria and measures, and (3) universal information management.

The first section on universal financial reports deals with the need for ag producers to standardize financial statements and reporting to provide for more accurate financial analysis of a given farming operation, facilitating the educational process for those interested in agriculture. As such uniform information is assembled over time, the development of benchmarks would lead to more attractively priced credit available to agricultural producers.

The second section deals with ratio analysis, which contributes to an understanding of the activities of the farm operation and enables management to measure the relative position of the operation within an industry group of similarly sized operations. Financial measures are not a substitute for informed judgment, so ratio analysis represents only a part of the overall evaluation that must occur in determining the success of a business.

The last section of the exposure draft discusses the development of a national database, through the use of an on-line computer system, that would include the development of standard industrial classification codes such as other industries have long enjoyed in making comparisons with industry averages, benchmarks, or both. The author gratefully acknowledges the efforts of the Farm Financial Standards Council, which continues to serve as a forum for problems and ideas relative to farm financial reporting and measurement.

Chapters 3 through 10 of this book discuss enterprise analysis, which is the process the producer undertakes to study the specific problems or opportunities that might be identified as a result of the financial and ratio analyses conducted in accordance with the first two chapters.

The figures shown throughout the book do not represent a specific producer, but they could be applicable to any producer. They are, in that sense, "real." I sincerely hope you will find this book informative and useful.

❧ How to Farm for Profit

1

Universal Financial Reports

OVER THE PAST EIGHT to 10 years, factors such as the explosion of the computer age, market volatility, and the farm debt crisis have heightened the demand for quality management information, including a move toward standardized financial reporting and financial analysis by agricultural producers. Leading the drive toward universal financial reporting has been the Farm Financial Standards Council (FFSC), formerly known as the Farm Financial Standards Task Force, which originated in January 1989 at a meeting of more than 50 volunteers representing various aspects of the agricultural finance community. Although the meeting was facilitated in large part by the Agricultural Bankers Division of the American Bankers Association, the FFSC evolved into an independent body whose initial efforts culminated in the issuance in May 1991 of *Recommendations of the Farm Financial Standards Task Force—Financial Guidelines for Agricultural Producers*. The council continued to meet and refine these recommendations, and in January 1995 the FFSC issued a revised publication in exposure draft form titled *Financial Guidelines for Agricultural Producers—Recommendations of the Farm Financial Standards Council*.

Our discussion in this chapter concerns financial reporting and is in large part based on the FFSC's recommendations. Al-

---

This chapter was written in collaboration with Michael H. Prosser, CPA.

though these recommendations are not required by regulation or are not otherwise authoritative within the realms of, for instance, financial reporting and credit file documentation, they do represent the consensus of opinion of professionals who have significant experience with or interest in the agricultural sector of our national economy. We wish to acknowledge the efforts put forth by these volunteers in tackling such an imposing project.

Uniform or standardized financial reporting is the label given to the data-gathering and assembling process that serves as a cornerstone for financial and enterprise analysis. Although ratio analysis (i.e., the FFSC's Sweet 16, discussed in the next chapter) has enjoyed higher visibility partly because it is less cumbersome and easier to understand than documented financial statements, it is logically appealing that the first subsection of the FFSC's recommendations deals with financial reporting as a step to be undertaken before one turns to ratio analysis.

The accounting profession has long adhered to a body of rules relative to the preparation and presentation of financial statements known as generally accepted accounting principles, or GAAP. The FFSC states in the introductory portion of its draft recommendations that it "believes that financial statements prepared in accordance with GAAP, together with certain supplemental information important to farmers and analysts, represents [*sic*] an appropriate standard for production agriculture." However, the FFSC acknowledges the shortcomings historically found in agricultural producers' record-keeping systems that made GAAP financial reports impractical to attain. In recognizing this problem, the FFSC stated that its report "does identify and provide suggested treatment for certain areas of financial reporting that, although not in accordance with GAAP, are in relatively widespread use among interested parties other than the accounting profession. In cases where farm businesses do not have and likely cannot afford to have the internal accounting systems necessary to generate financial statements in accordance with GAAP, the FFSC believes that the identified alternatives contained in this report, if adequately disclosed, still provide information useful to analytical purposes." In other words, the FFSC recognizes GAAP as the preexisting preferred method of accumulating and reporting financial data but allows for certain departures from GAAP in the interest of practicality and utility as long as such departures are identified in the financial presentation.

The FFSC recommends that the minimum set of financial statements should include (1) a balance sheet, (2) an income statement, (3) a statement of cash flows, and (4) a statement of owner's equity.

Financial reporting not only means financial statements but also includes footnotes and supplementary schedules considered necessary to add interpretation to and understanding of the financial condition of the agricultural producer. We will discuss these financial statements individually, but first we should review the basics of accrual basis accounting.

Accrual Basis Accounting

The FFSC endorses GAAP as the basis for farm financial statements that involve the use of accrual basis rather than cash basis accounting, yet only a small percentage of agricultural producers are using double-entry record-keeping systems that would accommodate the use of accrual basis accounting. In recognizing this situation, the FFSC recommends that an ag producer use the cash basis for recording the cash transactions occurring during a given accounting period (month, quarter, year) and then convert the data to accrual basis by making adjustments at the end of the period. The accounting profession has long touted the use of the accrual basis—matching revenues realized within the period against the expenses incurred necessary to generate such revenues—as the most representative measurement of the economic financial changes occurring within an accounting period. The crux of accrual basis accounting is that the actual date of receipt or payment of funds is irrelevant to the measurement of the results of operations. Instead, revenues are recorded when the sales occur (or in the case of readily marketable commodities, when the commodity is ready and available for sale), and expenses are recorded when incurred (i.e., when services or a product is received from a supplier) rather than when the bill for such an expense is actually paid. Because of the interplay between the balance sheet and the statement of income, the recording of inventories, accounts receivable, and accounts payable on the balance sheet generally results in an adjustment to an income or expense account in the statement of income so that the income statement is mostly converted from a cash to accrual basis through the process of the balance sheet accounts being adjusted to their proper end-of-period balances.

Let's look at a simple illustration. Assume the following facts about Farmer Jones:

1. Market value of harvested corn on hand at the beginning of the period: $20,000.
2. Market value of corn harvested during the current period: $100,000.
3. Cash sales of corn during the current period: $70,000.
4. Market value of harvested corn on hand at the end of the period: $50,000.

Assuming that no other transactions occurred during the accounting period, Farmer Jones would report net income on a cash basis of $70,000 because the cash sale of corn was the only cash transaction in our example. Accrual basis accounting ignores the amount of cash received when measuring the net income for the period and instead looks to the value of the corn harvested during the period ($100,000) as the accrual basis net income.

We note that the inventory of unsold grain has increased $30,000 since the beginning of the accounting period ($50,000 ending inventory minus $20,000 beginning inventory), so if we were to adjust the balance sheet inventory for the $30,000 increase, the "other side" of the entry would be a revenue account in the income statement. Where Farmer Jones is maintaining his records during the period on a cash basis, we would anticipate there being $70,000 in the grain sales/revenue account for the cash sales that occurred. It is interesting to note that the $30,000 inventory adjustment would serve to increase corn sales/revenues from $70,000 to $100,000, which is the accrual basis net income (equal to the value of the corn harvested).

To go a little bit farther, if Farmer Jones' corn inventory decreased from the start of the period to the end of the period, the adjusting entry would have been to decrease inventory on the balance sheet and likewise decrease revenue on the income statement. The purpose of this illustration is to show that the conversion from cash basis to accrual basis net income is simply a matter of taking cash basis revenues and expenses and increasing or decreasing them by changes in balance sheet accruals from the beginning of the period to the end of the period. The formula for converting from cash to accrual is as follows:

	Formula	Farmer Jones
Cash corn sales	$XXX	$70,000
Increase (decrease) in inventory of corn on hand	+/–XXX	+30,000
Accrual basis corn revenue	$XXX	$100,000

The same basic rule applies to accounts receivable, accounts payable, prepaid expenses, accrued interest payable, etc. The FFSC recommends that producers who are accustomed to using cash basis accounting continue to do so but then convert to accrual basis at the end of a given accounting period for the purposes of financial reporting. This conversion can be accomplished by developing the beginning and ending balance sheets that include all the necessary accruals, such as accounts receivable and accounts payable, and by increasing or decreasing cash basis income and expense accounts for the change in the accruals for the accounting period; thus we arrive at a converted income statement on an accrual basis. The accuracy of the accrual basis income statement depends on the accuracy of the beginning and ending balance sheets, so the producer should be diligent in determining the amounts to be recorded on them.

For example, assume that Farmer Jones wrote a check in December 19X1 for $10,000 to purchase seed to be used the upcoming spring but neglects to set up the $10,000 as a prepaid item on the December 19X1 balance sheet. Again, because Farmer Jones maintains his records on a cash basis, the $10,000 is sitting in seed expense at the end of the year; without a balance sheet adjustment for prepaid expense, this $10,000 will improperly be charged to 19X1 operations, when in fact it is going to be used in 19X2. Accrual basis accounting is the preferred method because items of revenue and expense are placed in the proper accounting period.

The Balance Sheet

Let's look at the purpose and composition of the four basic financial statements recommended by the FFSC to be presented in farm financial statements. First is the balance sheet. This statement is probably the most familiar to agricultural producers in large part because of the banking industry's practice of asset-based lending. Although a Schedule F from the producer's individual income tax return showing taxable farm income on a cash basis is an often-requested item to be placed in the credit file, most ag lenders concentrate their credit-making decisions on the balance sheet, or net worth statement, of the producer.

GAAP require that assets be recorded at their cost less an allowance for depreciation where applicable, yet rarely will an agricultural balance sheet report assets at their historical cost. Rather, a current market value approach will be used, which is not permitted by GAAP. Most nonagricultural industries carry their assets at historical cost, and lenders are accustomed to reviewing financial statements that have been prepared in conformity with GAAP. There are at least a couple of reasons for this use of current market value in farm financial reporting. First, much of the farmland has been held in the family for extended periods of time, and the original cost basis in the land may only be a fraction of its current market value. An ag lender would probably ask where the sense is in carrying farmland at $150 an acre when it currently would bring $2,000 an acre. Second, farm equipment, hog confinements, grain bins, etc., enjoy a rather rapid write-off under the income tax depreciation rules, and accordingly, their being recorded on a balance sheet at a cost less accumulated depreciation often results in a net book value significantly lower than their current market value. Again, a lender relying on long-lived or depreciable assets as collateral is interested in monitoring the current market value rather than historical cost, especially when the latter bears little relationship to the former. Another reason for using market value can be the unavailability of the cost basis information, especially when one is dealing with land, land improvements, or pieces of equipment that have been held for long periods of time.

Although the FFSC endorses GAAP, it recognizes the marginal value that cost information has for the typical agricultural producer and therefore condones the use of market values in the balance sheet; however, it also recommends that the historical cost information be disclosed either in a two-column format or in a schedule supplementary to the primary financial statements. Table 1.1 presents a two-column balance sheet with cost (book value figures) side by side with market value figures, both on an accrual basis, for a hypothetical farm producer. The cash basis column is shown for later use with our discussion of deferred income taxes and the income statement.

We see that receivables, inventories, and payables are recorded in both columns because accrual basis rules should be used irrespective of the valuation method. Although cost is normally the required valuation for GAAP, an exception allows inventories of readily marketable commodities to be valued at a net realizable value. We note that grain and livestock inventories in the GAAP column are recorded at their market values. A producer may find it difficult to determine the cost involved in harvested crops or growing animals. The primary dollar difference between book value and market value is contained in the noncurrent assets, as we might expect; real estate, machinery, and equipment have market values that are much higher than book values. Note that all liabilities are the same for book value and market value, which is the usual situation. In the owner's equity section, we seen an account labeled valuation equity that contains the amount representing the write-up of personal and noncurrent assets to their market values. Also in the equity section the line labeled current earnings represents the accrual basis net income, which is normally the same whether one is reporting on market value or book value terms. This is so because the differences between book value and market value are not captured in the statement of income but rather are included in the valuation equity account in the owner's equity section of the balance sheet.

Most farming enterprises are operated as proprietorships, many of which commingle personal assets—such as the fam-

Table 1.1. Balance sheet: December 31, 19X1

		Accrual basis	
Assets	Cash basis	GAAP book	Market
Current assets			
Checking account	$4,118	$4,118	$4,118
Accounts receivable	0	377	377
PIK certificates	0	1,889	1,889
Grain inventories	0	51,750	51,750
Livestock inventories	0	41,630	41,630
Total current	$4,118	$99,764	$99,764
Noncurrent assets			
Titled vehicles	$3,582	$3,582	$27,690
Machinery and equipment	105,229	105,229	296,570
Breeding livestock	84,129	84,129	170,330
Real estate	176,000	176,000	576,000
Building improvements	125,106	125,106	121,505
Total noncurrent	$494,046	$494,046	$1,192,095
Personal assets	1,420	1,420	2,400
Total assets	$499,584	$595,230	$1,294,259
Liabilities			
Current liabilities			
Notes payable	$359,960	$359,960	$359,960
Other payables	0	71,078	71,078
Accrued income taxes	0	40,903	40,903
Total current	$359,960	$471,941	$471,941
Noncurrent liabilities			
Equipment notes	$46,444	$46,444	$46,444
Farm mortgages	180,000	180,000	180,000
Total noncurrent	$226,444	$226,444	$226,444
Total liabilities	$586,404	$698,385	$698,385
Owner's equity			
Beginning capital	$(161,724)	$(96,649)	$(96,649)
Current earnings	102,259	20,847	20,847
Management withdrawals	(37,353)	(27,353)	(27,353)
Valuation equity	0	0	699,029
Total owner's equity	$(86,820)	$(103,155)	$595,874
Total liability and equity	$499,584	$595,230	$1,294,259

ily residence, nonfarm vehicles, savings accounts, etc., along with any related liabilities—with the farm assets and liabilities. This practice makes financial analysis more difficult to accomplish because it interferes with comparisons of ratios and relationships of other producers on an apples-to-apples basis. The FFSC recommends that if personal assets and liabilities are reported at all, they be shown separately after the farm assets and farm liabilities and be clearly labeled as personal so as to facilitate more meaningful financial analysis. Similarly, any nonfarm income should be reported in the statement of owner's equity as a contribution to the capital of the farm operation instead of being reported on the income statement.

Although the balance sheet of Table 1.1 has been adjusted to reflect accrual basis accounting and market values, a significant deficiency exists in the form of a failure to allow for the amount of income taxes that would be owed in the event of a hypothetical sale of a producer's assets at their accrued amount or market values—commonly referred to as *deferred taxes.* Deferred taxes are, in a sense, a contingent liability inasmuch as they are not owed to anybody as of the balance sheet date; however, they do tell the owners and creditors of the operation what part of the operation's assets would go for income taxes and therefore not be available for distribution to owners or payment to creditors should such assets be converted to cash in amounts exceeding their tax bases. Table 1.2 repeats the accrual GAAP and market value columns of Table 1.1, with the exception of recording a current liability and long-term liability for deferred taxes.

To further explain, the balance sheet shows real estate with a market value of $576,000, which is $400,000 over its cost basis of $176,000. The function of deferred taxes is to recognize the fact that should the real estate be sold at its market value, the resulting $400,000 gain would be subjected to income taxes of roughly one-third of such gain, or $133,333. This tells the owners and creditors of the farm operation that they should not count on using the full $576,000 of hypothetical proceeds from the sale of real estate at its market value for making payments to owners and creditors or otherwise use the cash in the farm operation because $133,333 will go for taxes. As we look at the owner's equity section of the balance sheet of Table 1.2, it is logically appealing that the equity has been decreased by the amount of deferred taxes when compared with the equity shown in Table 1.1.

Typically, two principal conditions create the need for recording of deferred taxes. First, most agricultural producers, especially those operating as partnerships and proprietorships,

Table 1.2. Balance sheet: December 31, 19X1, with liability for deferred taxes included

Assets	Market value	Market value with deferred taxes
Current assets		
Checking account	$4,118	$4,118
Accounts receivable	377	377
PIK certificates	1,889	1,889
Grain inventories	51,750	51,750
Livestock inventories	41,630	41,630
Total current	$99,764	$99,764
Noncurrent assets		
Titled vehicles	$27,690	$27,690
Machinery and equipment	296,570	296,570
Breeding livestock	170,330	170,330
Real estate	576,000	576,000
Building improvements	121,505	121,505
Total noncurrent	$1,192,095	$1,192,095
Personal assets	2,400	2,400
Total assets	$1,294,259	$1,294,259
Liabilities		
Current liabilities		
Notes payable	$359,960	$359,960
Other payables	71,078	71,078
Accrued income taxes	40,903	40,903
Current deferred taxes	0	9,828
Total current	$471,941	$481,769
Noncurrent liabilities		
Equipment notes	$46,444	$46,444
Farm mortgages	180,000	180,000
Long-term deferred taxes	0	223,689
Total noncurrent	$226,444	$450,133
Total liabilities	$698,385	$931,902
Owner's equity		
Beginning capital	$(96,649)	$(106,477)
Current earnings	20,847	20,847
Management withdrawals	(27,353)	(27,353)
Valuation equity	699,029	475,340
Total owner's equity	$595,874	$362,357
Total liability and equity	$1,294,259	$1,294,259

use the cash basis method of accounting for the purposes of filing their income tax returns, whereas for financial statement reporting purposes, they generally convert their cash basis records to an accrual basis, involving the setting up of receivables, inventories, and payables. These accrual adjustments affect the period in which revenues and expenses are to be reported, thus creating *timing differences.* These timing differences are used to calculate the current portion of deferred income taxes.

To illustrate, we refer back to Table 1.1, where the cash basis of grain and livestock inventories is zero, while on the accrual basis they total $93,380. By recognizing these inventories at their value rather than their tax basis, we have effectively added $93,380 to the net income of the current year. This "income" will not be reported until next year for tax purposes when the inventories are actually sold. The rules for deferred taxes require a liability to be recorded, recognizing the income taxes that will be payable when the assets having a book value higher than tax basis would be sold. Conversely, when payables are set up on an accrual basis, they represent expenses for the current period on an accrual basis but will not be deducted for tax purposes until the year they are actually paid, which would be the following year in most cases. This represents a negative timing difference that must be offset with all positive timing differences to determine the net timing difference to which the assumed marginal rate of income taxation must be applied to arrive at the current deferred taxes payable.

The second reason deferred income taxes are needed is that noncurrent assets are stated at their market values rather than at cost less an allowance for accumulated depreciation—also known as net book value. Because the noncurrent assets are not expected to be sold in the next accounting period, the deferred tax liability calculated on the valuation difference for noncurrent assets is considered to be a noncurrent liability. Accordingly, we can generalize by saying that there are two types of deferred taxes with which we must deal: current deferred taxes on the timing differences between the carrying value and tax basis of current assets and current liabilities, and long-term deferred taxes on the differences between the market value and book value of the noncurrent assets. It is also important to note that the change in current deferred income taxes is reflected in the provision for income taxes on the statement of income, thereby affecting net income for the period, while the change in long-term deferred income taxes is used to adjust the valuation equity component in the owner's equity statement and therefore does not affect current net income.

Although the concept of deferred income taxes can seem somewhat convoluted, we hope you will recognize that the purpose of deferred income taxes is to more accurately reflect the actual net worth of the agricultural producer by recognizing the potential income taxes payable in the future when assets are converted to cash. In Table 1.2, the owner's equity of $362,357 with deferred taxes recorded is more realistic than the $595,874 of equity reported on a straight market value basis without deferred taxes.

The Income Statement

We have reviewed the balance sheet and a few of the primary issues associated with it and during the course of the discussion have alluded to the income statement and its interplay with the balance sheet. Net income is an element of owner's equity that must be used to help balance the balance sheet. Additionally, converting a cash basis balance sheet to the accrual basis simultaneously converts a cash basis income statement to an accrual basis income statement. We therefore turn our attention to the income statement.

Most agricultural producers rely on Schedule F of their tax return as the only attempt to construct an income statement, and in some cases, they would not even complete Schedule F if it were not required by law. Schedule F can serve as a relatively accurate statement of cash income and cash expenses; however, only the taxable portion of some types of income and the deductible portion of some expenses are recorded on the schedule. More recently, some producers have moved to generating in-house income statements with their own computer and ag software, but even these are largely a collection of cash income statements that do not reflect any changes in accounts receivable, accounts payable, and inventory levels. Consequently, they are of limited value in understanding the real profitability of the farming operation. In fact, many of these in-house productions do not even include depreciation—a noncash expense included on every Schedule F. To include depreciation in an income statement during the course of an ongoing year before you know a year-end reconciled tax figure, you should use an estimated figure based on last year's final figure and adjust it at the end of the year before you close the year out. It allows you to think of, and evaluate your business, as a "whole operating entity" with all expenses accounted for at any given time during the course of the year.

Using the same example illustrated above for the balance sheet, let's review the income statement shown in Table 1.3. For the most part, the cash basis column represents income items received in cash and expense items paid in cash during the year. Because of this approach, a producer's decisions about what products to sell in December versus January or which expenses to pay before or after year end can cause the net income on a cash basis to fluctuate widely even though the actual "true" income for each accounting period is fairly stable. Quite typically, income tax factors dictate the timing of cash transactions, thereby making the cash method unpredictable as a measurement of profitability. Unfortunately for the producer illustrated in Table 1.3, the accrual basis income of $20,847 is probably a more accurate measurement of profits than the cash basis net income of $102,257.

If we look closely at the differences between the two columns in Table 1.3, the revenues from crops and from market livestock account for the difference in gross revenues and quite likely represent increases or decreases in inventory levels and not more or less economic production for the year. Operating expenses are only a couple of thousand dollars different on the cash basis versus the accrual basis, but interest expense is almost $24,000 higher on the accrual basis. Here the producer probably did not pay up his/her interest with the bank at the end of the year to the same extent as he/she had done in the previous year; so again, this is not indicative of more profit but rather is the *impression* of profitability resulting from a conscious decision by the producer as to the timing of the interest payment. The potential misstatement of profits by using a cash basis is vividly illustrated by the provision for income taxes, which is reported as zero on the cash basis. You might at first wonder how zero income taxes on a net income of $102,257 can be proper reporting; however, keep in mind that on the cash basis only those income taxes actually paid during the year ending December 31, 19X1, would be reported as a cash basis expense. Chances are good in the first four

Table 1.3. Income statement for the year ended December 31, 19X1

	Cash basis	Accrual basis
Gross revenues		
Gross revenue from crops	$163,054	$202,893
Gross revenue from market livestock	342,876	257,063
Livestock products	118,117	118,117
Government programs	106,332	106,332
Other farm income	20,254	20,254
Total gross revenues	$750,633	$704,659
Less cost of purchased livestock	113,200	113,200
Less cost of other purchases sold	12,440	12,440
Total purchases	$125,640	$125,640
Value of farm production	$624,993	$579,019
Operating expenses		
Feed purchases	$53,481	$50,418
Fertilizer, lime, chemicals	44,069	44,069
Freight, trucking	36,352	36,352
Gas, fuel, oil	15,738	15,738
Insurance (not including crop insurance)	4,791	4,791
Labor hired	29,394	29,394
Machinery hired	4,527	4,527
Land rent	76,164	76,164
Repair and maintenance	27,941	27,941
Seeds and plants	13,997	13,997
Supplies	6,734	6,734
Taxes	9,633	10,320
Utilities	8,575	8,575
Vet, medical, breeding	3,721	3,721
Crop insurance, irrigation, drying	34,347	34,347
Farm business expense	9,138	9,138
Depreciation expense	71,288	71,288
Miscellaneous	7,387	7,387
Total operating expenses	$457,277	$454,901
Interest expense	62,208	86,120
Total expenses	$519,485	$541,021
Net farm income from operations	$105,508	$37,998
Gain (loss) on sale of assets	(3,251)	(3,251)
Net farm income before income taxes	$102,257	$34,747
Provision for income taxes	0	13,900
Net income	$102,257	$20,847

months of the following year that we will see a big cash basis deduction for income taxes paid about the time the 19X1 tax returns actually get mailed to the government. This is a prime example of the failure of the cash basis of accounting to properly match revenues with expenses.

What was the cause of the difference in the accrual basis versus the cash basis figures? We can refer to Table 1.4 for the

Table 1.4. Items on income statement converted from cash to accrual basis for the year ended December 31, 19X1

		Conversion items, increase (decrease), cash basis income
Cash revenue from crops	$163,054	
+/– Change in inventories	37,950	$37,950
+/– Change in PIK certificates	1,889	1,889
Gross revenue from crops, accrual basis	$202,893	
Cash revenue from market livestock	$342,876	
+/– Change in inventory	(86,190)	(86,190)
+/– Change in accounts receivable	377	377
Gross revenue from market livestock, accrual basis	$257,063	
Cash basis operating expenses	$457,277	
+/– Change in accounts payable	(3,063)	3,063
+/– Change in taxes payable	687	(687)
Operating expenses, accrual basis	$454,901	
Cash interest paid	$62,208	
+/– Change in accrued interest	23,912	(23,912)
Interest expense, accrual basis	$86,120	
Provision for income taxes: cash paid	$0	
+/– Change in current portion of deferred taxes	(27,003)	27,003
+/– Change in accrued income tax	40,903	(40,903)
Provision for income taxes, accrual basis	$13,900	
Net effect of conversion items		$(81,410)
Cash basis farm net income		102,257
Accrual basis farm net income		$20,847

answer. As we mentioned above, converting the balance sheet from cash basis to accrual basis by establishing receivables, inventories, and payables simultaneously converts the income statement from a cash to accrual basis. By taking cash basis income and expenses and adjusting for changes in these accrual accounts on the balance sheet from the beginning to the end of the period, we achieve the desired conversion effect.

The provision for income taxes on an accrual basis of $13,900 warrants further discussion. Table 1.4 shows that the provision for income taxes is composed of two numbers: a decrease in the current deferred taxes and an increase in accrued income taxes. We already know that the purpose of the deferred taxes is to recognize the timing differences between cash basis and accrual basis reporting, which is generally limited to changes in current assets and current liabilities. By taking the accrual conversion items from Table 1.4, we can arrive at the net change in such items of ($67,510).

Increase in crop and inventory	$37,950
Increase in PIK certificates	1,889
Decrease in market livestock inventory	(86,190)
Increase in accounts receivable	377
Decrease in accounts payable—feed	3,063
Increase in taxes payable	(687)
Increase in accrued interest payable	(23,912)
Net timing differences	$(67,510)
Assumed marginal rate of income tax	×40%
Change in current portion of deferred taxes	$(27,003)

Because the net timing difference is a negative figure (meaning that the differences between current assets and current liabilities recorded at their cost basis when compared with the accrual basis are smaller at the end of the year than the difference at the beginning of the year) the amount of current deferred taxes needs to be recorded to reflect the difference between accrual basis and cash basis. This, however, is only one component included in the provision for income taxes.

Accrued income taxes are different from deferred income taxes inasmuch as they represent taxes that are actually owed on income reported to the taxing authorities through the balance sheet date (we don't owe deferred taxes to anyone yet). In our example financial statements, our producer reported a cash (tax) basis net income of $102,257 (see Table 1.3); with an assumed rate of taxation of 40%, the amount of income taxes payable with respect to this net income would be $40,903. Because beginning-of-the-year accrued income taxes were zero, the change in accrued income taxes payable would be the $40,903. Although the payment of these taxes will not occur until early the following year, remember that the accrual basis requires the recording of receivables and payables that exist as of the balance sheet date (December 31, 19X1).

Referring back to Table 1.1, we note that there are no accrued income taxes payable in the cash basis column; however, the $40,903 accrual is recognized for GAAP purposes as well as the market value method. This simply is another example of how, by converting the balance sheet from a cash to accrual basis, the income statement is forced to go to accrual basis at the same time. To summarize, it is important to remember that the provision of income taxes on an accrual basis income statement is made up of the total (or net) of the change in current deferred income taxes and accrued income taxes. To test the accuracy of the provision for income taxes calculated above, we simply can take the net farm income before taxes on an accrual basis as shown on Table 1.3 in the amount of $34,747 times the 40% marginal income tax rate assumed, and we arrive at the same $13,900 provision for income taxes.

The FFSC has some more recommendations concerning the handling of items on the income statement. In brief, the more important are the following:

1. Unlike the situation in some other industries, agricultural producers are not expected to calculate a "cost of goods sold" or "gross profit" but rather should derive a subtotal called "value of farm production," or VFP, which represents gross revenues less the cost of purchased livestock, feedstuffs, or other items for resale. The reason for using the VFP is to avoid the distortions that can happen when large purchases of items for resale occur toward the end of an accounting period. For example, a cattle feeder could purchase $250,000 of livestock for resale in December, but because revenues from feeder cattle are adjusted for changes in inventory levels, revenues would be artificially overstated by such a purchasing pattern. Although this overstatement is offset by the increase in livestock purchases, so net income remains unchanged, any ratios tied to gross revenues would be distorted. Use of cost of goods sold is discouraged because of the difficulty in determining direct expenses and allocating indirect expenses to the cost of production. VFP is limited to deducting only the purchases of items for resale, which is more simple to quantify.

2. No expense should be recorded for unpaid labor and management provided by the owners. Net farm income should be viewed as the return not only on capital but also on owner labor. Owner withdrawals/family living expenses should be reported in the statement of owner's equity and not as an expense on the income statement.

3. Gains and losses from sales of breeding livestock and equipment should be included in the calculation of net farm income rather as an extraordinary item reported at the bottom of the income statement. Although these transactions may be fairly infrequent in nature, they still occur in the normal course

of business, and the FFSC has concluded that they properly belong in the calculation of net farm income.

4. Use of income tax methods of depreciation are permissible when the aggregate amount charged to operations for the year does not differ materially from the depreciation expense calculated with GAAP rules. This problem has become less important in recent years as the Internal Revenue Code has become more restrictive in the accelerated rate at which depreciable assets can be written off, but the $17,500 annual Section 179 deduction still remains and serves to overstate depreciation expense; it should be avoided for financial reporting purposes where it is material in relationship to the income statement taken as a whole.

5. We previously discussed the exception to the GAAP/accrual basis rule requiring assets to be recorded at cost for commodity inventories that can be valued at market when they are readily available and intended to be converted to cash in a short period of time. The FFSC recommends that when raised crop inventories are intended to be fed to livestock rather than sold, they should be recorded at below their cost or market value—the argument being that the market valuation method runs the risk of distorting income by reflecting unrealized gains on products that are not intended to be sold in their present form.

The Statement of Cash Flows

We now move to the third of the four basic financial statements, the statement of cash flows. Traditionally, a statement of cash flows has been viewed as a projection of future transactions, most frequently on a monthly basis, to demonstrate debt repayment capacity to a producer's lender. The statement of cash flows contemplated by the FFSC is the GAAP statement, a sample of which is shown in Table 1.5. This statement is a historical statement covering the same period as the income statement and segregates an entity's cash flows into three main sections—operating activities, investing activities, and financing activities. With this statement, the focus shifts from profits to cash flows. The bottom line of a cash flow statement represents the increase or decrease in cash for the

Table 1.5. Statement of cash flows for the year ended December 31, 19X1

		Accrual basis
Cash flows from operating activities		
Net income (loss)		$20,847
Adjust to reconcile net income (loss) to		
net cash provided by operations:		
Depreciation	71,288	
Loss (gain) from sale of assets	3,251	
Decrease (increase) in receivables and PIK certificates	(2,266)	
Decrease (increase) in inventory	48,240	
Increase (decrease) in accounts payable	(3,063)	
Increase (decrease) in accrued interest due	23,912	
Increase (decrease) in taxes payable	687	
Increase (decrease) in accrued income tax	40,903	
Increase (decrease) in current deferred income tax	(27,003)	
Total adjustments		155,949
Net cash provided by (used in) operations		176,796
Cash flows from investing activities		
Proceeds from sales—noncurrent assets	14,924	
Payment for purchase—noncurrent assets	(82,540)	
Proceeds from sales—personal assets	0	
Payment for purchase—personal assets	(1,420)	
Net cash provided by (used in) investing		(69,036)
Cash flows from financing activities		
Proceeds from operating loans	928,122	
Proceeds from noncurrent loans	28,960	
Principal paid on operating loans	(1,015,727)	
Principal paid on noncurrent loans	(24,913)	
Dividends/owner draw/family living	(27,353)	
Net cash provided (used in) financing		(110,911)
Net increase (decrease) in cash/equivalents		(3,151)
Cash/equivalents at beginning of period		7,269
Cash/equivalents at end of period		$4,118
Supplemental disclosure of cash information		
Interest paid	$62,208	
Income taxes	$0	

year. By also showing the beginning-of-the-year and end-of-the-year cash coming from the producer's balance sheet, we can be sure that all cash flows have been accounted for.

The operating activities section starts with the accrual basis net income from the income statement because net income is considered the primary source of cash from operational activities. However, because it is an accrual basis figure, we must undo the effects of the accruals by adding or subtracting the changes from the beginning to the end of the year in all the

other current asset and current liability accounts to revert back
to a cash basis number. Additionally, depreciation and loss or
gain from the sale of assets need to be added back because in
the first instance we are dealing with an expense that did not
involve an outlay of cash, and in the second instance the cash
flow from the sale of equipment is reported in the investing activities section. We note in Table 1.3 that although the first column is a cash basis presentation of the income statement, it
does include depreciation and loss on the sale of farm capital
assets, both of which did not involve an outlay of cash and accordingly, for cash flow purposes, should be added back to our
cash basis net income of $102,257.

Net income—cash basis	$102,257
Depreciation—add back	71,288
Loss on sale of assets—add back	3,251
Net cash provided by operations, per statement of cash flows	$176,796

As we can see, the statement of cash flows takes virtually
the reverse approach from what we did to develop the accrual
basis income statement from cash basis information. Instead,
we are starting with accrual basis net income and removing all
the accrual adjustments to get back to a cash basis number.

The investing activities section in Table 1.5 is reserved for
purchases and sales of noncurrent assets such as real estate and
depreciable property, purchases and sales of investments, and
purchases and sales of personal nonfarm assets. Table 1.5
shows our example producer using $69,036 of cash in invest-
ing activities primarily for the acquisition of depreciable assets
net of the proceeds from the sale of such assets.

The section on financing activities is used for recording proceeds received from new borrowings as well as principal payments on existing borrowings; these numbers should be shown
at gross rather than net payments against proceeds. Although it
can be argued that interest expense is part of financing activities, it is an item that runs through the income statement in determining net income and therefore must be in the operating
activities section of the cash flow statement. The other type of
transactions properly reported in the financing section are any
contributions of cash from the owners and any payments made

to the owners for family living, owner draws, etc. Table 1.5 shows that $110,911 of cash was used in financing activities, as principal payments on loans exceeded proceeds from new loans, and the owner drew $27,353 for family living.

To summarize, our producer generated $176,796 of cash flow from operations (income statement), used $69,036 in investing activities for purchasing equipment, and used $110,911 in financing activities making loan principal payments and owner draws. The net increase or decrease in cash cannot be construed as a measure of profitability, efficiency, or any other benchmark of success. It simply reflects the change in the balance of cash during the year. An increase in cash is not necessarily good nor is a decrease in cash necessarily bad, unless the decrease reduces the ending cash balance to a negative number, which might be a problem in the banker's mind. Our producer of Table 1.5 is fairly typical in that the enterprise's operations were the primary source of cash, which the producer then used for investing in new capital assets and/or making debt payments and distributions to him/herself. The statement of cash flows serves as the answer to the owner's question, "If the income statement shows I made $20,847 during the year, then why do I have less cash in the bank now than I did at the beginning of the year?"

The Statement of Owner's Equity

The fourth and final of the standardized financial statements is the statement of owner's equity. Owner's equity is what's left when liabilities are subtracted from assets on the balance sheet; therefore, this statement is crucial because it assures the preparer of the financial statements that the accrual adjustments and assets and liabilities have all been properly recorded. The statement of owner's equity is needed to tie the beginning balance sheet to the ending balance sheet.

As shown in Table 1.6, the statement of owner's equity starts with the beginning-of-the-year equity, which is the sum of the beginning contributed capital and retained earnings and the beginning valuation equity account. The first section of this statement includes the net income or loss for the year, owner's withdrawals, and any contributions of capital (includ-

Table 1.6. Statement of owner's equity for the year ended December 31, 19X1

Beginning capital (deficit)		$(106,477)
Beginning valuation equity		414,277
Owner's equity beginning of period		$307,800
Net income (loss)	20,847	
Management (family living) withdrawals	(27,353)	
Capital contributions/gifts/inheritance	0	
Capital distributions/gifts/dividends	0	
Total change in contributed capital and retained earnings		(6,506)
Change in excess of market value over cost/basis	88,818	
of farm noncurrent assets		
Change in value of personal assets	980	
Change in noncurrent portion of deferred taxes	(28,735)	
Total change in valuation equity		61,063
Owner's equity end of period		$362,357

ing items of nonfarm income), which are added together to arrive at the subtotal for change in contributed capital and retained earnings. Next, the changes in the valuation equity account are reported. These changes include the increase or decrease of the excess market value over the cost basis of the noncurrent assets, change in value of any personal assets, and change in the noncurrent portion of deferred taxes, which is tied to the change in market values as we have previously discussed. The changes in these two sections are then added to the beginning equity number to arrive at the ending equity number, which must agree with the equity shown on the balance sheet of $362,357, using the market values with deferred taxes column of Table 1.2.

We have covered the four primary financial statements recommended by the FFSC in its section on universal financial reports, but you must remember that GAAP and the FFSC recommendations also require footnote disclosures and supplementary schedules to be included with the four primary statements whenever such information is necessary for the understanding of the producer's financial condition. The financial statements and related notes serve as the building blocks from which financial and enterprise analysis can be performed.

Financial Analysis, Ratios, and Indicators

FINANCIAL ANALYSIS, particularly the use of selected financial ratios, has been accepted as a method of measuring financial position and performance. Farming enterprises tend to be more difficult to analyze than other industry types because of the variety in the size of agricultural producers as well as the degree of diversification that might be found within the operations of a particular producer. These factors, along with the lack of standardized financial reporting found within the agricultural industry, prompted the Farm Financial Standards Council (FFSC) to formulate recommendations regarding the selection and use of financial measures and ratios. The FFSC identified the following criteria as important in agricultural financial analysis:

1. *Liquidity* measures the ability of the farm business to meet financial obligations as they come due in the ordinary course of business, without disrupting the normal operations of the business.
2. *Solvency* measures the amount of borrowed capital (or debt), leasing commitments, and other expense obligations a business uses relative to the amount of owner's equity invested in the business. Debt capital is interest bearing and/or has a date by which it must be paid. There-

fore, solvency measures provide (a) an indication of the ability of the business to repay all financial obligations if all assets were sold (for the prices indicated), and (b) an indication of the ability of the business to continue viable operations after a financial adversity (such as drought), which typically results in increased debt or reduced equity.

3. *Profitability* measures the extent to which a business generates a profit from the use of land, labor, management, and capital.

4. *Repayment capacity* measures the ability of the borrower to repay term farm debt from farm and nonfarm income. Principal payments on term loans must come from net income (with depreciation added back) after owner withdrawals, income taxes, and social security taxes.

5. *Financial efficiency* measures the intensity with which a business uses its assets to generate gross revenues and the effectiveness of production, purchasing, pricing, financing, and marketing decisions.

The goal of the FFSC was to arrive at a list of the most commonly used financial measures that would best inform the analyst as to the financial position and performance of the ag producer in the five areas listed above. Although financial ratio analysis has long been the domain of agribusiness lenders as a systematic approach to credit analysis, the informed owner/manager should learn how to interpret the financial signals that such analysis can yield.

The Sweet 16 Recommended Farm Financial Measures

The FFSC has developed a list of financial measures, frequently referred to as the Sweet 16, that are felt to provide the most meaningful financial and management information. The Sweet 16 list, along with formulas for the calculation of each item, follows.

Liquidity
1. Current Ratio=Total Current Farm Assets/Total Current Farm Liabilities

2. Working Capital=Total Current Farm Assets–Total Current Farm Liabilities

Solvency

3. Debt/Asset Ratio=Total Farm Liabilities/Total Farm Assets
4. Equity/Asset Ratio=Total Farm Equity/Total Farm Assets
5. Debt/Equity (leverage) Ratio=Total Farm Liabilities/Total Farm Equity

Profitability

6. Rate of Return on Farm Assets=(NFIFO[1]+Farm Interest Expense–Value of Operator and Unpaid Family Labor and Management[2])/Average Total Farm Assets[3]
7. Rate of Return on Farm Equity=(NFIFO–Value of Operator and Unpaid Family Labor and Management)/Average Total Farm Equity
8. Operating Profit Margin Ratio=(NFIFO+Farm Interest Expense–Value of Operator and Unpaid Family Labor and Management)/Gross Revenue
9. Pretax Net Farm Income (from income statement)

Repayment capacity

10. Term Debt and Capital Lease Coverage Ratio= (NFIFO+Total Nonfarm Income+Depreciation Expense+Interest on Term Debt and Capital Leases–Total Income Tax–Withdrawal for Family Living)/Annual Scheduled Principal and Interest Payments
11. Capital Replacement and Term Debt Repayment Margin=NFIFO+Total Nonfarm Income+Depreciation Expense–Total Income Tax–Family Living Withdrawals–Payment on Unpaid Operating Debt From Prior Period–Principal Payments on Current Portion of Term

---

[1] NFIFO=Net farm income from operations after interest expense but before noncurrent asset sales and income taxes.

[2] Owner withdrawals are often used, or an estimated value of the owner's labor factor can be used.

[3] Beginning-of-year balances added to end-of-year balances divided by two is commonly used. Monthly averages for the year produce a more accurate measure but are more difficult to compute.

Debt and Capital Leases–Total Annual Payments on Personal Liabilities (if not included in family living withdrawals)

Financial efficiency
12. Asset Turnover Ratio=Gross Revenue/Average Total Farm Assets
13. Operating Expense Ratio=Operating Expenses/Gross Revenue
14. Depreciation Expense Ratio=Depreciation Expenses/Gross Revenue
15. Interest Expense Ratio=Interest Expenses/Gross Revenue
16. Net Farm Income From Operations Ratio=NFIFO/Gross Revenue

Although these financial measures are simply a matter of mathematics, their usefulness to the analyst is limited by the quality of the financial statements from which the information is derived. Rigid reliance on these financial measures for evaluation of financial performance can be misleading. Knowledge of certain nonfinancial factors and informed judgment are vital parts of meaningful financial analysis.

A Hypothetical Analysis

Let's use the financial statement information from the tables in Chapter 1 to plug into the Sweet 16 financial measures and see what conclusions we can draw regarding our hypothetical producer's financial position and financial performance for 19X1.

Liquidity

Measure and equation	Values	Result
1. Current ratio		
Current assets	$99,764	
÷Current liabilities	÷482,769	=0.21
2. Working capital		
Current assets	$99,764	
−Current liabilities	−482,769	=$−382,005

COMMENTS: *There's a big problem here. Any current ratio less than 1.0 indicates an inability to pay debts when they come due. The notes payable probably are, for the most part, lines of credit that although "come due" annually, are routinely extended by the lender. If not, debt restructuring will be required.*

Solvency

Measure and equation	Values	Result
3. Debt/asset ratio		
Total farm liabilities	$931,902	
÷Total farm assets	÷1,294,259	=0.72
4. Equity/asset ratio		
Total farm equity	$362,357	
÷Total farm assets	÷1,294,259	=0.28
5. Debt/equity ratio		
Total farm liabilities	$931,902	
÷Total farm equity	÷362,357	=2.57

COMMENTS: *Here, there's too much debt. Any equity/asset ratio less than 0.40 is a danger signal. Interest expense could wipe out profits. Note that these three ratios are all algebraically interrelated and simply represent different approaches to reporting solvency.*

Profitability

Measure and equation	Values		Result
6. Rate of return on farm assets			
NFIFO	$37,998		
+Farm interest expense	86,120		
−Value of operator labor	(27,353)		
	$96,765		
÷Average total farm assets		$\dfrac{\$96,765}{\$1,276,674}$	=7.58%
Beginning-of-year	$1,259,089		
End-of-year	1,294,259		
	$2,553,348		
	÷2		
	$1,276,674		
7. Rate of return on farm equity			
NFIFO	$37,998		
−Value of operator labor	(27,353)		
	$10,645		
÷Average total farm equity		$\dfrac{\$10,645}{\$335,079}$	=3.18%
Beginning-of-year	$307,800		
End-of-year	362,357		
	$670,157		
	÷2		
	$335,079		
8. Operating profit margin ratio			
NFIFO	$37,998		
+Farm interest expense	86,120		
−Value of operator labor	(27,353)		
	$96,765		
÷Gross revenues		$\dfrac{\$96,765}{\$704,659}$	=13.7%
9. Net farm income			=$34,747

COMMENTS: *Return on assets is above average, but this is calculated without interest expense in the picture. Interest costs could be unmanageable. The value of operator labor may be low, which tends to show better rates of return.*

Repayment capacity

Measure and equation	Values	Result
10. Term debt and capital lease coverage ratio		
NFIFO	$37,998	
+Total nonfarm income	0	
+Depreciation expense	71,288	
+Interest on term debt	36,120	
−Income taxes	(13,900)	
−Owner withdrawals	(27,353)	
	$104,153	
+Annual scheduled principal and interest payments		
Mortgage payments	$27,000	
Equipment note payments	25,000	
	$52,000	$\dfrac{\$104,153}{\$52,000}$ =2.00:1
11. Capital replacement and term debt replacement margin		
NFIFO	$37,998	
+Total nonfarm income	0	
+Depreciation expense	71,288	
−Income taxes	(13,900)	
−Family living withdrawals	(27,353)	
−Payments on prior period operating debt	0	
−Principal payments: current portion of term debt	(15,880)	
−Total payments on personal liabilities	0	=$52,153

COMMENTS: *A coverage ratio of 1.0 is the minimum needed to make the scheduled payments. Anything above 1.5 is good. This producer has a significant amount of short-term debt with no scheduled payments but is at risk that the lender will not renew the line of credit. A restructuring of debt to schedule payments out into the future is a consideration. The purpose of the replacement margin is to measure the producer's ability to generate funds necessary to replace capital assets and make debt payments beyond the current period.*

Financial efficiency

Measure and equation	Values	Result
12. Asset turnover ratio		
Gross revenue	$704,659	
÷Average total assets	÷1,276,674	=0.55
13. Operating expense ratio		
Operating expense	$383,613*	
÷Gross revenue	÷704,659	=0.54

(*Operating expenses should be reduced by depreciation: Total Operating Expense [$454,901] − Depreciation [$71,288] = $383,613.)

14. Depreciation expense ratio		
Depreciation	$71,288	
÷Gross revenue	÷704,659	=0.10
15. Interest expense ratio		
Interest	$86,120	
÷Gross revenue	÷704,659	=0.12
16. Net farm income from operations ratio		
NFIFO	$37,998	
÷Gross revenue	÷704,659	=0.05

COMMENTS: *Operating expenses of less than 65% of gross revenues is considered good. Depreciation is about normal, and interest expense ratios of between 0.10 and 0.20 are not troubling, but still we have enough interest here that it reduces return on assets from 7.58% to 0.83%. The asset turnover ratio is tough to gauge, but the ag industry in general shows poorly in this area because of the capital-intensive nature of operations and partly because of the use of market value for farmland in the denominator of the turnover ratio. A recommended target for the NFIFO ratio is 10% to 15%. Our producer would be right there if he/she could reduce the cost of borrowing.*

Financial Ratio Analysis and Profitability

A good understanding of your business begins with an examination of financial liquidity and solvency, which are directly influenced daily by profitability. This examination is generally referred to as *financial ratio analysis*. Many lenders are now trying to use ratios to analyze and compare loan customers, one with another, in order to evaluate the strength of customers and their total loan portfolio. Herein lies the difficulty. Very few benchmarks for ag ratio interpretation have been published, and loan officers are left on their own to interpret the ratios and your financial strength. Therefore, understanding your ratios, and how they are being interpreted, very quickly becomes your business. After all, your ratios and how they are compared with others may mean the difference between money and no money.

One of the problems in interpreting ratios is the diverse nature of agriculture. Different types of farming and ranching operations legitimately produce differing ratios. When your operation is being compared with another, make sure you know who you are being compared with. To illustrate, Table 2.1 compares five distinct types of farming operations: a cash grain farming operation, a diversified grain and livestock farming operation, an operation of only livestock, a cow/calf ranch, and a commercial cattle feedlot. Notice that an acceptable current ratio, debt/asset ratio, or return on equity in one type of operation is not acceptable in another. Yet very few people understand the nature of examining agriculture. This is why it is incumbent upon you to know your type of operation and the appropriate ratios.

The value of using financial ratios for analysis is twofold. Ratios provide a common denominator relationship that allows for ease of multiyear analysis. Also, combining information

Table 2.1. Comparison of five types of farming operations

Ratios and indicators	Cash grain	Grain and livestock	Livestock only	Cow/calf	Commercial feedlot
Liquidity					
Current ratio	2.5:1	2.0:1	1.75:1	2.5:1	1.5:1
Solvency					
Debt/asset percent	40%	50%	60%	35%	65%
Debt/equity ratio	0.67:1	1.0:1	1.5:1	0.54:1	1.86:1
Profitability					
Return on assets	3%	7.5%	5.5%	2.5%	4.4%
Return on equity	5%	15.0%	13.75%	3.85%	12.6%
Financial efficiency					
Asset turnover ratio	0.35	0.80	1.5	0.25	1.75

from the balance sheet and income statement to produce ratios allows you to understand how the statements influence each other. To illustrate, let's examine four years of ratios from a hypothetical diversified grain/livestock farming operation. The ratios used in Table 2.1 are typical of a successfully managed grain/livestock operation and are repeated in the first column of Table 2.2.

The current ratio illustrates the number of dollars of current assets existing for each $1 of current debt. A current ratio of 2.0:1 is listed in column 1, indicating that $2 of current assets exist for each $1 of current debt. Therefore, the operator could cash in adequate current assets to pay off all current debts and still have one-half of all current assets remaining. A review of the current ratio illustrates a decrease from 2.0:1 to 1.4:1 in the four-year period. Therefore, we can reason that the operator is losing his/her ability to pay all current debts without interfering with fixed (long-term) assets and the uninterrupted continuation of normal operations.

The solvency factors illustrate an increasing percentage of debt each year and an increasing amount of debt for each dollar of equity. Both constitute negative trends. This immediately becomes a concern of debt lenders and should concern the operator as well. The question is: Why is this happening? Return on assets decreased from 7.5% to a negative 6.0%, and return on equity decreased from 15.0% to a negative 12.0%. This operation is not profitable.

An important lesson is that the value of the balance sheet is

Table 2.2. Four years of ratios for diversified grain/livestock operation

Ratios and indicators	Column (year)			
	1	2	3	4
Liquidity				
Current ratio	2.0:1	1.8:1	1.7:1	1.4:1
Solvency				
Debt/asset percent	50%	52%	53%	56%
Debt/equity ratio	1.0:1	1.08:1	1.13:1	1.27:1
Profitability				
Return on assets	7.5%	1.2%	0.53%	(6.0%)
Return on equity	15.0%	0.6%	(3.0%)	(12.0%)
Financial efficiency				
Asset turnover ratio	0.80	0.75	0.75	0.65

directly and irrevocably tied to the success (profitability) of the income statement—not the other way around.

There are many reasons for unprofitable operations, and ratio analysis cannot fully explain any of them. However, the asset turnover ratio does give some clues in this example. The operator originally turned (produced) $0.80 of gross annual income for each $1 of total assets. This reduced to 0.65:1 and may give a hint of future problems.

Everyone in business should examine all the financial and operating ratios of the business at least once a year and in so doing should ask some questions: Is my business profitable? Is the balance sheet strong and growing stronger? These are questions we must all ask ourselves if we are going to survive into the 21st century. Originally, we all enter a business or profession because we like doing whatever it is that the job requires. Only after committing ourselves to our chosen lifetime work do we begin to understand how important it is to be profitable (to make money) in order to stay in business.

Unfortunately, in agriculture that hasn't always been true. During the years of unbridled inflation, ag producers and lenders alike became comfortable with increasing the values for real estate and equipment annually when constructing the business balance sheet. In effect, inflation made everyone *look* profitable. However, be assured that *looking* profitable is not the same as *being* profitable.

The press would have us believe that many hundreds of farmers went out of business during the 1980s because asset values fell. In other words, it's the fault of asset evaluation. But that's not the entire truth. Farmers went broke because they no longer *looked* profitable and didn't know how to *become* profitable. Profitability had become an unimportant factor of doing business.

As stated above, the balance sheet is irrevocably tied to the income statement, not the other way around. Unless you're altering asset values, the balance sheet cannot look profitable and equity cannot increase without actually producing a profit. When a business produces a profit, it shows up on the bottom line of the income statement.

We make this point for two reasons. First, you cannot understand the value of ratio analysis or financial statements until you accept this basic premise of business life. And to stay in business and be profitable, it's important to understand statements and analyses. Second, some lenders are quietly returning to the habit of increasing asset values annually, and many farmers are willing collaborators.

So why are people back to changing asset valuations every year? Because people aren't as profitable as they would like to appear . . . because it makes life easy . . . because you get something for nothing. This may be true, but it's not real and not permanent.

During the 1980s, 34% of farmers left the business, and during the 1990s, we are on schedule to lose another 30%. It's time for agricultural producers to become realistic and make their businesses profitable—and they will actually discover that it's enjoyable to be in business when making a profit.

Ratios: Just the Beginning

As we've been discussing, financial analysis of a farming or ranching operation begins with ratio analysis—a very important exercise. Ratio analysis ties the value of the balance sheet and income statement together and allows you to begin understanding your operation as a business. Most important, if you use all of the FFSC's Sweet 16 ratios, the analysis will reveal any existing operational problems. I stress the use of all 16 ratios because among users (including lenders) little exists today that resembles a standardized approach. Some use as few as two or three ratios to examine business stability. An example is to examine only the liquidity and solvency ratios of the balance sheet. Although you may get a good feeling about the balance sheet, you barely have a clue as to the profitability of operations.

Just as important, you should recognize that although financial ratios inform you of an operational problem, ratios will not identify the problem or suggest how to solve it. Please remember, operational problems are the root cause of profitability problems, so it's extremely important not only to recognize such problems but also to solve them. Although this may sound rudimentary, most operating problems go unsolved, not because management doesn't have the desire to do so, but because the problem cannot properly be identified. The proper identification and solving of operational problems require more than ratio analysis. Financial ratios are just the beginning. As an example, we have looked at ratios in an operation of declining profitability, coupled with a declining asset turnover ratio. A declining asset turnover ratio attests to the fact that (today) fewer dollars of gross income are being produced for each $1 of assets used in the operation. A learned reader will quickly conclude that a decreasing use of assets is in process. Although this ratio is an accurate predictor of potentially less-profitable operations, one is left wondering why the ratio is down and what to do about it.

This, of course, is the difficult part. It's easy to tell the producer "you've got a problem." Now it's up to the producer to find the problem and solve it. That's what management is all about, and as margins are reduced, management becomes more important.

The proper identification and solving of operational problems are left to enterprise (profit center) analysis, which we'll begin to explore in the next chapter. Financial analysis involves the study of "financial facts"—balance sheet and income statement facts. Enterprise analysis involves the study and interrelationship of financial facts and "operational facts." This is where the fun starts.

ᘒᓓ 3
Enterprise (Profit Center) Analysis

AS IMPORTANT AS FINANCIAL ratios are, they are just the beginning of a complete and informative financial analysis of your business. Enterprise analysis is an important follow-up to understanding ratios. Why? Without enterprise analysis, you cannot understand the break-even cost of doing business, that is, the cost of production.

Enterprise analysis requires taking all operating expenses and breaking them down into the lowest common denominator of production. This means, for instance, expenses per bushel, per ton, or per hundredweight of production. The difficulty is that most farming and ranching operations have several enterprises with a variety of different profit centers, such as cropping, cow/calf production, cattle and hog feeding, turkey production, and so on. Additionally, some major enterprises may have subcategories, such as corn, soybeans, and wheat, within the cropping enterprise.

Therefore, the question becomes, How do you correctly divide expenses into the different enterprises and subcategories? A direct input cost is not very hard. For example, most fertilizers and chemicals are allocated to the cropping enterprise. But what about the indirect and overhead expenses? How do you divide management and labor costs, depreciation, taxes, and other costs into different enterprises? The answer is that you must form a methodology for dividing all expenses among enterprises. And importantly, the method you use must remain constant from one year to the next.

Following are some suggestions for formatting the separation of expenses in a farming operation. Remember, you will not be able to undertake enterprise analysis until you do this.

Operating Expenses: Method of Separation

When you are prepared to take all operating expenses for a year and separate them as illustrated below, the process of enterprise analysis can proceed. To begin, we recommend that you apply the procedure to several past years. Then, you can begin allocating expenses within a current accounting system.

Management/labor:	(a) Complete daily time log or
	(b) Use an hours/unit formula.
Interest:	(a) Purchase cost calculation or
	(b) Dollars/unit formula.
Depreciation:	Calculate equipment depreciation and divide into enterprises based on percentage of use.
Repair/maintenance:	Calculate equipment original cost and apply a percentage of total to actual repair and maintenance costs.
Fuel/oil:	(a) Daily time log or
	(b) Hours/unit formula.
Supplies:	(a) Same as fuel/oil.
Insurance:	(a) Employee/liability: Same as management/labor;
	(b) Vehicle/livestock: same as repair/maintenance or direct allocation.
Utilities/other:	(a) Same as management/labor or
	(b) Direct allocation.
Rent, feed, machine hire, seeds/plants, fertilizer/chemicals, vet/ medical, and storage:	Direct allocation or dollars/unit formula.

Help With Expense Breakdown

To help you further with this division of expenses among different enterprises, let's look more closely at the breakdown of expenses for management/labor.

The cost of hired labor is easy to calculate. A check is normally written for this expense, and adding all of the *gross* payroll together equals an annual total. How does one calculate the cost of management, recognizing that management is normally performed by the owner (often a husband/wife team), with checks written in amounts necessary to cover family living expenses? If money is available for family living expenses that provides a reasonable lifestyle (say, $20,000 to $35,000 a year), call that the cost of management. If you're unhappy with that, you can use a formula, such as an assumed salary level of $30,000 plus 10% of net pretax profits.

Assuming a total management/labor cost of $90,000, how should it be divided? One way is with a formula approach, as shown in Table 3.1. In very large operations it is important for employees to account for the time they spend in each enterprise or task. By using the formula of Table 3.1, you can see how management and labor time can be allocated to an enterprise without hourly time justification. And it is accurate, assuming that the annual hours per unit have been regionally adjusted.

Now you can allocate the $90,000 total cost of management/labor to each enterprise as shown in Table 3.2. A cautionary note: If certain employees are part time or seasonal, allocate all of that labor cost to the enterprise involved.

Table 3.1. Dividing management/labor costs

Enterprise	Hours/unit	Total units	Total hours employed Crops	Cattle	Hogs	Cow/calf
Corn	3.2	300	960			
Soybeans	2.8	150	420			
Alfalfa	4.0	100	400			
Cattle feeding	3.0	2,000		6,000		
Hogs (farrow-to-finish)	15.0	250			3,750	
Cow/calf	8.0	300				2,400
Total hours in operation			1,780	6,000	3,750	2,400
Percent			13%	43%	27%	17%

Table 3.2. Allocation of management/labor costs by enterprise

	Crops	Cattle	Hogs	Cow/calf
Percentage of $90,000	13%	43%	27%	17%
Enterprise total	$11,700	$38,700	$24,300	$15,300

Management/labor cost is a "leader" line-item expense. The expense makes a statement about the separation of all expenses and can be used as a guide to separate other expenses, such as taxes, supplies, insurance, utilities, and miscellaneous.

It's important to recognize that allocating expenses is a step-by-step procedure with multiple calculations, and you must pay attention to detail. Computers and good software have made the process easy and fast.

Enterprise Expenses

Having formatted a methodology for segregating expenses among the different enterprises of a farming and ranching operation, we're now ready to spread the expenses for a given year or years. Remember, you must be prepared to divide *all* expenses of the operation, not just direct input cost. Select a year (or more) of operating expenses to work with. These expenses can be taken from Schedule F of your tax return, an internal record-keeping system, or an accounting statement. Having listed each expense, it is wise to divide the expenses by a percentage of use so as not to lose any dollars. The results of your effort should resemble Table 3.3, and completing the expense separation by percentage should result in the expenses for each enterprise shown in Table 3.4.

After you finish the process shown in Tables 3.3 and 3.4, you can add up all of the expenses in each enterprise for your grand total. What is the benefit in knowing this? The benefit is knowing the complete cost of each enterprise, not just the *direct* cost, but the *total* cost, including overhead, operations, and depreciation. The benefit is knowing how much money you spent, for instance, on raising crops versus the different categories of livestock involved.

Once you've divided your expenses among enterprises, you know the cost of each one and can successfully tell your lender

Table 3.3. Categories of expenses and percentage to each enterprise

Expense	Amount	Percentage of amount to				
		Crops	Cattle	Hogs	Dairy	Cow/calf
Management/labor	$90,000	27	15	5	34	19
Fertilizer/chemicals	51,000	90	0	0	5	5
Trucking	21,000	0	25	15	55	5
Fuel/oil	15,600	41	24	10	19	6
Insurance	7,800	41	14	7	30	8
Machine hire	5,500	100	0	0	0	0
Rent	54,500	65	0	0	0	35
Repair/maintenance	27,000	52	8	5	28	7
Seed/plants	14,750	100	0	0	0	0
Supplies	6,500	35	12	8	40	5
Taxes/utilities	9,600	27	15	5	34	19
Vet/medical	5,800	0	20	7	38	35
Interest	86,000	38	16	9	20	17
Depreciation	72,000	45	13	10	20	12

"I know that cattle feeding, hogs, etc. is not where my problem lies."

Common Denominators

We've used a formatted method for segregating all expenses into the different enterprises of a farming and ranching operation. The result is that the total operational expense of each enterprise is known. This process has begun to make the information of an income statement useable for day-to-day management decisions. The importance of this cannot be underestimated. An income statement contains much good information that is normally hidden unless one knows how to extract it. Enterprise expenses information is the beginning. But to be completely useable, the enterprise expenses information must be coupled with one more prerequisite: common denominators.

Common denominators allow you to understand (break down) the total enterprise expenses on a per-unit basis. It's important to select the right common denominators: those that provide management information. Here are some suggestions, using the most common enterprises as examples.

1. Cropping: (a) dollars per acre; (b) dollars per unit harvested (per bushel per hundredweight per ton). To calcu-

Table 3.4. Expenses per enterprise

Expense	Crops	Cattle	Hogs	Dairy	Cow/calf
Management/labor	$24,300	$13,500	$4,500	$30,600	$17,100
Fertilizer/chemicals	45,900	0	0	2,550	2,550
Trucking	0	5,250	3,150	11,550	1,050
Fuel/oil	6,396	3,744	1,560	2,964	936
Insurance	3,198	1,092	546	2,340	624
Machine hire	5,500	0	0	0	0
Rent	35,425	0	0	0	19,075
Repair/maintenance	14,040	2,160	1,350	7,560	1,890
Seed/plants	14,750	0	0	0	0
Supplies	2,275	780	520	2,600	325
Taxes/utilities	2,592	1,440	480	3,264	1,824
Vet/medical	0	1,160	406	2,204	2,030
Interest	32,680	13,760	7,740	17,200	14,620
Depreciation	32,400	9,360	7,200	14,400	8,640

late, you must know the total acres cropped, acres per individual crop, and units harvested per acre.

2. Livestock feeding (cattle, hogs, sheep, turkeys, etc.): (a) dollars per head fed; (b) cents per head day; (c) dollars per hundredweight (per pound) of gain. To calculate, you must know the starting weight in, the ending weight out, the number of head fed, and the total days on feed.

3. Dairy: (a) dollars per cow (lactating and gestating); (b) dollars per milk cow (lactating only); (c) dollars per hundredweight of milk produced (sold). To calculate, you must know the total number of cows in the herd, the number milking, and the hundredweight of milk produced.

4. Cow/calf: (a) dollars per bred cow in herd; (b) dollars per calf weaned; (c) dollars per hundredweight of calves weaned. To calculate, you must know the total number of bred cows in the herd, the number of calves, and hundredweight weaned.

Take some time to select your own common denominators for your farming enterprises. Once you select them for each of your enterprises, you are ready to apply them to the enterprise expenses. Next, we'll show you the results of applying the expenses to common denominators in our example case. This is where the information starts to become meaningful. Remember, the information has to be useful to management.

Cropping Enterprise Analysis

HAVING DISCUSSED the necessary prerequisites of enterprise analysis (determining enterprise expenses and selecting common denominators) and split our operational expenses into cropping, cattle feeding, hog feeding, dairy, and cow/calf enterprises, we're now ready to analyze the different enterprises in our example farming and ranching operation, beginning with the cropping enterprise.

Cropping Expenses

The cropping operations consist of 1,000 acres of ground, some rented and some owned, cropped one-third each to corn, soybeans, and alfalfa. Table 4.1 shows the total dollar amount of each expense as it was allocated to the cropping enterprise. The second column illustrates an average cost per acre if the expense was evenly distributed over the 1,000 acres. However, in actuality, some crops will cost more or less than others so the remaining columns illustrate a split of expenses for each individual crop. Notice that the expenses are divided into fixed and variable expenses. The division of expenses is important to sensitivity analysis ("what-if" games). Sensitizing crops is a function of forecasting future operations, which will be discussed in a later book.

Table 4.1. Cropping enterprise expenses

Expense	Total	Per acre	Corn	Soybeans	Alfalfa
Fixed expenses					
Management/labor	$24,300	$24.30	$23.80	$20.82	$28.25
Insurance	3,198	3.20	7.00	1.30	1.30
Rent	35,425	35.42	53.19	53.19	0.00
Taxes/utilities	2,592	2.59	2.59	2.59	2.59
Depreciation	32,400	32.40	38.00	38.00	21.23
Total	$97,915	$97.91	$124.58	$115.90	$53.37
Variable expenses					
Fertilizer/chemicals	$45,900	$45.90	$75.00	$45.50	$17.29
Fuel/oil	6,396	6.40	6.00	5.52	7.60
Machine hire	5,500	5.50	8.26	8.26	0.00
Repair/maintenance	14,040	14.04	16.25	15.75	10.13
Seed/plants	14,750	14.75	23.00	13.07	8.20
Supplies	2,275	2.27	2.79	2.79	1.25
Interest	32,680	32.68	36.08	33.48	28.50
Total	$121,541	$121.54	$167.38	$124.37	$72.97
Total expenses	$219,456	$219.45	$291.96	$240.27	$126.34

Now let's carry the analysis a step further. Let's repeat the total expenses line and assume that corn produced 120 bushels per acre, soybeans produced 35 bushels per acre, and alfalfa produced 4 tons per acre. To find a break-even cost of production, divide each crop cost by the production per acre, as shown in Table 4.2.

Only once in the past 10 years has the average cash price of corn been in excess of the $2.43 breakeven shown in Table 4.2 (when it was $2.54 per bushel in 1988/89). The same is true for soybeans ($6.86 in our example; $7.42 per bushel in 1988/89). Alfalfa produced at $31.58 per ton is always profitable.

Other than for a potential profit on alfalfa hay, the only net income generated from these crops is U.S. Department of Agriculture set-aside payments. When a situation like this

Table 4.2. Break-even costs of production for each crop

	Corn	Soybeans	Alfalfa
Total cost/acre	$291.96	$240.27	$126.34
Production/acre	120 bu	35 bu	4 tons
Corn cost/bu	$2.43		
Soybean cost/bu		$6.86	
Alfalfa cost/ton			$31.58

arises, additional analysis should include breaking all crops down into owned, cash rented, and crop share ground. Table 4.3 shows a cost allocation illustration for the corn crop. You can and should use the same procedure on every crop within the cropping operation.

Notice that the cash rent paid is $80 per acre. The crop share ground illustrated is a one-third to two-thirds sharing, reducing the yield, seed, and fertilizer/chemical expense. The taxes and interest are reduced in both rentals. Now let's find a breakeven for each method of operation, as shown in Table 4.4. You'll notice that the spread of break-even costs among the three operating methods is somewhat typical although many producers have experienced greater spreads between owned and rented ground—sometimes as much as $0.50 per bushel. Ground is often bid up and rented above its equivalent ownership value. However, much of the farm ground sold is purchased in the same manner—beyond its ability to produce a profit.

Breaking down corn crops between owned and rented ground shows that although there is a modest difference, it is not significant. As an example, dropping cash rented ground will not significantly reduce the cost. Therefore, there must be other problems related to all acres cropped.

Once you break down the analysis of cropping operations

Table 4.3. Corn cropping expenses broken down by owned, rented, and shared land

Expense	Corn average	Owned	Cash rent	Crop share
Fixed expenses				
Management/labor	$23.80	$23.80	$23.80	$23.80
Insurance	7.00	7.00	7.00	7.00
Rent	53.19	0.00	80.00	0.00
Taxes/utilities	2.59	9.49	1.39	1.39
Depreciation	38.00	38.00	38.00	38.00
Total	$124.58	$78.29	$150.19	$70.19
Variable expenses				
Fertilizer/chemicals	$75.00	$80.00	$80.00	$49.95
Fuel/oil	6.00	6.00	6.00	6.00
Machine hire	8.26	8.26	8.26	8.26
Repair/maintenance	16.25	16.25	16.25	16.25
Seed/plants	23.00	26.00	26.00	15.32
Supplies	2.79	2.79	2.79	2.79
Interest	36.08	69.98	19.50	18.75
Total	$167.38	$209.28	$158.80	$117.32
Total expenses	$291.96	$287.57	$308.99	$187.51

Table 4.4. Break-even costs by method of operation

	Owned	Cash rent	Crop share
Total cost/acre	$287.57	$308.99	$187.51
Production/acre (bushels)	120	120	80
Break-even cost/bu	$2.40	$2.57	$2.34

into individual major crops, and owned versus cash rented versus crop share rented, sensitivity analysis should also examine each field or tract of land. The differences from one to another are interesting. In our example the differences between owned and rented ground were not significant, demonstrating that any problems were related to all acres cropped. To further analyze, let's examine each line of expenses (overall per acre average expenses) and apply the amount to a benchmark spending level (derived from Agri Control data bank, USDA information services, and Doane information services).

1. Management/labor ($24.30 per acre): This cost should normally not exceed $18 to $20 per acre. Potential savings: $4.30 per acre.
2. Insurance ($3.20 per acre): This is relatively in line as an overall average. It will be crop specific and will range from $2 to $10 per acre.
3. Rent ($35.42 per acre): Rent must be evaluated for individual fields. When renting at $80 per acre, if crop breakeven is high when compared with owned ground, the rent may be high.
4. Taxes/utilities ($2.59 per acre): This value is relatively in line.
5. Depreciation ($32.40 per acre): This is much too high for nonirrigated ground and should average $20 to $22 per acre. There's too much equipment sitting around. Potential savings are $10.40 per acre with better utilization.
6. Fertilizer/chemicals ($45.90 per acre): This value is crop specific.
7. Seed/plants ($14.75 per acre): This must be analyzed by crop and will range widely, depending on crop agronomy.

8. Fuel/oil ($6.40 per acre): This is well within normal anticipated levels.
9. Machine hire ($5.50 per acre): A norm for this expense cannot be anticipated except as it affects the total machinery cost per acre.
10. Repair/maintenance ($14.04 per acre): This is $3 to $4 per acre too high.
11. Supplies ($2.27 per acre): This value is relatively in line.
12. Interest ($32.68 per acre): This value is carrying approximately $120 per acre equipment debt and $600 per acre land debt. It's relatively high for 120 bushels per acre corn ground. Better utilization or sale of equipment could save $7.50 per acre in interest.

Note that the most troublesome expenses are management/labor, depreciation, rent, repair/maintenance, and interest, all of which are related to management and land and equipment costs. Everyone thinks of direct input costs such as fertilizers, chemicals, and seed as being too high, but these costs are generally not the primary cause of unprofitable cropping operations. Instead, it's fixed overhead costs. The good news is—these are expenses management can do something about.

If one were to reduce management/labor by $4.30 per acre, depreciation by $10.40 per acre, repair/maintenance by $3.00 per acre, and interest by $7.50 per acre, the total would be a savings of $25.20 per acre. This reduces the breakeven for producing corn by $0.21 per bushel, soybeans by $0.72 per bushel, and alfalfa by $6.30 per ton. With these adjustments, corn would cost $2.22 per bushel; soybeans, $6.14 per bushel; and alfalfa, $25.28 per ton.

Solving the Cropping Problem

The crops in our example are simply not making any money. Although one could save $25.20 per acre by reducing costs to the benchmark "not-to-exceed" figures suggested above, the difference is not enough to make cropping profitable. Let's examine what would happen if we sold equipment, rented 500 acres more land (crop share), or did 500 acres of custom crop-ping (Table 4.5). The costs must be reduced, and the fixed costs plus interest are the most important.

Table 4.5 shows that enough equipment was sold to reduce depreciation by one-half. But this does not reduce any other fixed expenses, and machine hire might go up. The cropping of additional ground is a better overall choice because it re-duces all fixed costs plus interest if rented and all except rent plus interest if custom cropped. This action still has not re-duced soybeans to the point of normal profitability.

Beyond this, one may have to change the cropping plan and rotation to become more profitable, but we're making progress now. The point is to examine your own operation and readjust until crops can be produced at a profit. Until then, cropping is an exercise in futility. Most advertising encourages you to spend more money, to grow more per acre, or to buy more horsepower to become more efficient. But what's most impor-

Table 4.5. Cropping expenses after selling equipment, renting land, and custom cropping

Expense	Present cost/acre	Sell equipment	Rent land (500 acres)	Custom crop 500 acres
Fixed expenses				
Management/labor	$24.30	$24.30	$16.20	$16.20
Insurance	3.20	3.20	2.13	2.13
Rent	35.42	35.42	23.62	35.42
Taxes/utilities	2.59	2.59	1.73	1.73
Depreciation	32.40	16.20	21.60	21.60
Total	$97.91	$81.71	$65.28	$77.08
Variable expenses				
Fertilizer/chemicals	$45.90	$45.90	$45.90	$45.90
Fuel/oil	6.40	5.40	6.40	6.40
Machine hire	5.50	11.50	5.50	5.50
Repair/maintenance	14.04	9.00	14.04	14.04
Seed/plants	14.75	14.75	14.75	14.75
Supplies	2.27	2.27	2.27	2.27
Interest	32.68	27.68	21.79	17.75
Total	$121.54	$116.50	$110.65	$106.61
Total expenses	$219.45	$198.21	$175.93	$183.69
Reduction per acre		$21.24	$43.52	$35.77
Resulting crop cost				
Corn (120 bu/acre)	$2.43	$2.25	$2.07	$2.13
Soybeans (35 bu/acre)	6.86	6.26	5.62	5.84
Alfalfa (4 tons/acre)	31.58	26.27	20.70	22.64

tant is becoming more efficient by better utilizing assets. That's the key.

Specialty Crops: A Separate Enterprise

To this point we have looked at only the most common crops in terms of total population and gross national receipts. There are many other cropping specialties; however, one person's specialty crop might be common to another. For our discussion we'll define specialty crops as fruits and vegetables, versus drilled/row crop grains (which are specialty crops for many farmers in the Midwest).

During the 1980s, when it appeared that everyone was going broke raising corn and soybeans, agricultural agencies and other supposedly informed pundits were encouraging farmers to raise "specialty" or alternative crops—as though the *crop* was the problem. If one isn't profitable raising crops common to an area, why would one expect to be profitable raising something never tried before?

Nonetheless, for those already into specialty crops, either solely or in conjunction with row crops, or thinking about entering this realm, the field does include special problems concerning agronomy, financing, and marketing. We'll look at the

Table 4.6. Cost per acre of production for grain acres and vegetables

Expense	Corn	Soybeans	Cabbage	Sweet corn	Beans	Potatoes
Fixed expenses						
Management/labor	$22	$22	$61	$40	$28	$42
Depreciation	23	23	320	163	23	66
Other	10	10	6	6	10	10
Total	$55	$55	$387	$209	$61	$118
Variable expenses						
Labor	$0	$0	$295	$87	$0	$219
Fertilizer/chemicals	86	30	145	216	99	270
Freight/supplies/marketing	5	5	249	125	9	88
Fuel/oil	15	14	200	54	19	554
Machine hire	8	8	6	21	115	8
Repair/maintenance	17	17	225	62	17	103
Seed/plants	29	22	140	56	110	99
Utilities	5	5	185	38	5	15
Other	34	24	178	96	82	148
Total	$199	$125	$1,623	$755	$456	$1,504
Total expenses	$254	$180	$2,010	$964	$517	$1,622

financial aspects. The cost of growing a specialty crop is very different from that of corn or soybeans and very different among many specialty crops. This is best illustrated by the figures of a producer involved in the production of both grain acres and vegetables (Table 4.6).

Do not use these figures to calculate your operation. Calculations vary from one operation to another. The cost variances shown in Table 4.6 dictate that row crops and specialty crops should be treated as separate enterprises.

Also, if the specialty crops are marketed directly in a retail facility, the retail operation should be treated as a separate enterprise, purchasing all products from the farm, therein creating an income to the farm and a cost-of-sales charge to the retail operation. Without this transfer, one can't make an informed decision about an adequate retail price.

Cattle Enterprise Analysis

NEXT LET'S EXAMINE the cattle feeding enterprise of our hypothetical producer. To examine this enterprise we must break down the expenses into dollars per head fed, cents per head day, and dollars per hundredweight of gain. Actually, cents per head day is a good common denominator for assessment of feedlot efficiency and dollars per hundredweight for evaluation of cattle performance.

Feedlot Cattle Expenses

Our first objective is to see how the feedlot works (Table 5.1). The feedlot holds 300 head, creating a maximum capacity of 109,500 cattle-on-feed days (300 head×365 days per year), but it's illustrated at 99,000 days, or about 90% of maximum capacity (calculation: 2.2 annual turns×300 head×150 days=99,000 days). The expenses are broken down per head and per head day.

In addition to total fixed and variable operating expenses, we must add total feed cost (Table 5.2). Feed is cost specific to the kind of cattle being fed. In this example we are using 750-pound feeders being finished to a 1,250-pound weight and will assume a daily average feed cost of $1.03 per day and an average daily gain (ADG) of 3.3 pounds per day.

Table 5.1. Feedlot cattle expenses: fixed and variable

Expense	Total	Per head	Per head day
Fixed expenses			
Management/labor	$13,500	$20.45	$0.136
Insurance	1,092	1.65	0.011
Taxes/utilities	1,440	2.18	0.015
Depreciation	9,360	14.18	0.095
Total	$25,392	$38.46	$0.257
Variable expenses			
Trucking	$5,250	$7.95	$0.053
Fuel/oil	3,744	5.67	0.038
Repair/maintenance	2,160	3.27	0.022
Supplies	780	1.18	0.008
Vet/medical	1,160	1.76	0.012
Interest	13,760	20.85	0.139
Total	$26,854	$40.68	$0.272

Table 5.2. Cattle operating expenses plus feed costs

	Total	Per head	Per head day
Total operating costs	$52,246	$79.14	$0.529
Total feed costs	101,970	154.50	1.030
Total all costs	$154,216	$233.64	$1.559
Cost/lb gain (ADG of 3.3 lb)			$0.472

Our profit/loss calculation looks like this:

Purchase cost	$547.50 (750 lb × 0.73/lb)
Cost of gain	$233.64 (from above)
Sales price	$818.75 (1,250 lb × 0.6550/lb)
Profit per head	$37.61

Having completed the analysis, one can see that if the feedlot is operated at full capacity (90% of max), and is achieving a 3.3 average daily gain, a profit can be made at today's purchase and selling prices. The questions which management of every operation should ask are: "What are my operating costs per head day?" "Can I achieve a 2.8 ADG?" Or better yet, "Is the feedlot full?"

As always, the importance of management and the efficient use of assets is uppermost in determining success. Next we'll take a look at some variations in asset use.

Less-Than-Full Feedlot

We have examined the feedlot at 90% of maximum capacity (99,000 cattle-on-feed days), which is essentially full given the fact the individual pens cannot be filled with new feeders on the same day that finished cattle leave. As such, the fixed and variable expenses are amortized at $.472 per head day. Some improvement could be made in Mgmt/Labor, Depreciation, and Fuel/Oil costs. But the lot is profitable, largely because of a good ADG (3.3) and a full lot.

Many don't realize how important it is "to be full." And the reasons for not keeping the pens full range from not cleaning the pens out fast enough, no financing from the banker, to not wanting cattle when in the field harvesting. Some fully equipped feedlots operate as low as 40% of capacity. The problem is that "Fixed Expenses" are fixed, and continue on regardless of whether cattle are on feed or not. The expenses are just divided between fewer cattle. This has a tremendously negative effect on profits. The chart illustrates the cost of fixed expenses at 90%, 75%, 60%, and 45% of capacity.

Let's look at what happens to the per head day expenses in our cattle enterprise analysis if the feedlot is less than full, at 45%, 60%, 75%, and 90% of capacity (Table 5.3).

A few cents per head day doesn't seem like much money.

Table 5.3. Feedlot cattle expenses at less than feedlot capacity

Expense	Total	Per head day expenses			
		90%	75%	60%	45%
Management/labor	$13,500	$0.136	$0.164	$0.205	$0.274
Insurance	1,092	0.011	0.013	0.017	0.022
Taxes/utilities	1,440	0.015	0.018	0.022	0.029
Depreciation	9,360	0.095	0.114	0.142	0.190
Total fixed	$25,392	$0.257	$0.309	$0.386	$0.515
Total variable	26,854	0.272	0.272	0.272	0.272
Total operating	$52,246	$0.529	$0.581	$0.658	$0.787
Plus feed cost		1.030	1.030	1.030	1.030
Total daily maintenance cost		$1.559	$1.611	$1.688	$1.817
×150 days on feed		$232.85	$241.65	$253.20	$272.55
Additional cost per head		$0	$7.80	$19.35	$38.70
Total cattle sold		600	547	438	328
Total loss on cattle sold		$0	$4,267	$8,475	$12,694

But as you can see from Table 5.3, when you multiply by 150 days on feed, the dollars start adding up. When this hypothetical operation drops from 90% to 75% of capacity, $7.80 of cost per head fed is added to the cattle. A producer can drop to 75% capacity just by waiting 20 days to fill the pen. And the loss shown in Table 5.3 does not include the opportunity lost on cattle not fed.

These figures can help you understand that management is the most important ingredient to the operation of a profitable feedlot. Remember, in this example, everything seems to be running smoothly: the feed rations are good, the cattle are gaining 3.3 pounds per day, and the cattle are bought and sold right; however, the feedlot isn't full. Good management practices begin with understanding the cost of doing business. Only after knowing that the normal cost is $0.472 per head day can one calculate the cost of not having a full feedlot or the cost of not marketing correctly.

Some feedlot managers (especially those in larger feedlots) would disagree with how we placed some of our fixed expenses, such as labor and utility costs, arguing that both should be considered variable expenses; and indeed they can be. However, over time I've discovered that many expenses normally regarded as variable in nature become fixed in small operations. For instance, once the hired man steps out of his pickup, he's there for the day, regardless of how many cattle are on feed or what the weather is like.

That's why it's so easy for small operations to become inefficient, and the reason why small operators must be doubly cautious about the full utilization of assets. Many times I have stated publicly that the small operator can successfully compete next door to a very large operation. However, both must be *efficient*, and both must *fully utilize assets*. This is equally difficult for both operators, which is why strong management is equally important. Otherwise, neither will survive.

When we examined the effect of not fully utilizing an asset, that is, not keeping the feedlot full, we illustrated the negative effect on profits of having to divide all fixed expenses among fewer than 600 head of cattle fed during the year. At 75% of capacity, 547 head were fed and sold and had to absorb an additional $4,267 of fixed expenses; at 60% capacity, 438 head absorbed an additional $8,475; and at 45%, 328 head absorbed $12,694. However, this is only half of the story. If 547, 438, and 328 head of cattle were sold instead of 600 head, then a profit opportunity was also lost on 53, 162, or 272 head not fed.

We originally charted the profit objective at $37.61 per head fed and sold at 1,250 pounds. In carrying this objective forward, we can say that the cattle "not sold," because there was no opportunity to feed them, would equal an additional loss of profits as follows:

Percentage of capacity	90%	75%	60%	45%
No. of cattle not fed	0	53	162	272
Profits lost ($37.61/head)	$0	$1,993	$6,093	$10,230

The whole story includes a potential loss incurred because cattle fed were required to amortize excess fixed expenses, and opportunity profits were lost on those not fed.

	Percentage of capacity			
	90%	75%	60%	45%
Cattle sold	$0	$4,267	$8,475	$12,694
Cattle not sold	0	1,993	6,093	10,230
Total losses	$0	$6,260	$14,568	$22,924

The original profit forecasted for the feedlot was $22,566 (600 head×$37.61 per head). At 75% of capacity the feedlot will lose more than 25% of the original objective. At 60% the loss is more than 66% of the original, and at 45% the entire profit objective is lost—and management might as well stay home.

It seems that at times the cattle feeding industry is too profitable because so many operators feel they can feed cattle if and when they want to. That's not true. If you're in, you're in 100% of the time. If not, you'll soon be 100% out.

Daily Gain and Cattle Profits

We've been examining the negative effects on profits produced by a feedlot operated at less-than-full capacity. An even more insidious thief of good profitability is the lack of good daily gains.

Most managers subscribe to a minimum level of ration analysis, usually provided by a feed company. But beyond that most are far too casual about how the ration is mixed, the daily feeding times, bunk and water cleanliness, and a host of other factors that affect daily weight gain. For instance, many managers weigh the first potload of cattle sold out of a pen (44 to 46 head), calculate the gain, and brag about the 3.2 pounds per head day. But how about the last load at 1.9 pounds per day? The days—and dollars—required to finish a steer rise dramatically as daily gain decreases:

Average daily gain (lb)	3.3	3.1	2.8	2.6
Days on feed required	150	162	175	191

Table 5.4. Expense charges to cattle and daily gain

	Total	Dollars/day	Dollars/head			
Fixed expenses	$25,392	$0.257	$38.46	$41.42	$44.86	$48.95
Variable expenses						
Trucking	5,250	...	7.95	7.95	7.95	7.95
Fuel/oil	...	0.038	5.67	6.16	6.65	7.26
Repair/maintenance	...	0.022	3.27	3.56	3.85	4.20
Supplies	...	0.008	1.18	1.30	1.40	1.53
Vet/medical	...	0.012	1.76	1.94	2.10	2.29
Interest	...	0.139	20.85	22.52	24.32	26.55
Total expenses			$79.14	$84.85	$91.13	$98.73
Per hundredweight of gain			$18.84	$20.20	$21.70	$23.51

Again, as you can see in Table 5.4, the fixed expenses are absorbed by fewer cattle on feed longer. But this time the variables also get involved, remaining the same on a head-per-day basis and increasing in total when the number of days increases. Only a couple of minor expenses, such as trucking and marketing, are exempt.

We haven't examined the feed cost yet, and some argue that the feed cost per pound of gain decreases if daily gain achieved is lower. I will disagree. As an example, let's use a static cost of $36.79 per hundredweight of gain, or $154.51 per head fed and sold. Assuming a purchase cost of $550 per head, the results are as shown in Table 5.5.

Table 5.5. Gains (losses) after feed costs

	Days			
	150	162	175	191
Total cost of gain	$233.65	$239.36	$245.66	$253.26
Purchase cost	550.00	550.00	550.00	550.00
Total of all costs	$783.65	$789.36	$795.66	$803.26
Break-even sales price/cwt	$65.30	$65.78	$66.30	$66.94
Loss per head sold	$0	$5.76	$12.00	$19.68

The potential loss is insidious. It's just a few cents each day, until the dollars are added up. Remember, if cattle are taking longer to finish, fewer are being fed. Our original feedlot could finish 660 head per year at 150 days on feed, 611 head at 162 days, 565 head at 175 days, and 518 head at 191 days. Without calculating the opportunity lost on cattle not fed, the above loss is $3,519 on 611 head, $6,780 on 565 head, and $10,194 on 518 head fed and sold. How many times would you have liked to have shown your banker another $10,000 of profit?

6

Hog Enterprise Analysis

WE HAVE EXAMINED the cropping and cattle feeding enterprises of our hypothetical producer. The third enterprise, hogs, is a totally confined farrow-to-finish profit center consisting of 60 breeding female hogs, boars, and finishing hogs. The 60 female hogs are bred twice annually in groups of 30, each generating four farrowings per year, and produce 7.5 weaned pigs per litter (15 pigs annually) for a total of 900 weaned pigs. After a 2.2% death loss, 880 feeders are finished to 240 pounds annually. (Incidentally, the term weaned pig is the most often used vernacular and in this case identifies a 40-pound feeder coming out of the nursery.)

Breeding and Finishing Subenterprises

To properly analyze a farrow-to-finish operation, it's important to separate all breeding herd and hog finishing expenses, in effect creating two subenterprises. If you don't do this, it's impossible to understand and evaluate each piece of the operation. For example, knowing how many pigs are weaned is important but of little value without knowing how much they cost to wean. You can't calculate this cost when the expenses are combined with the costs of finishing hogs.

Review the example shown in Table 6.1 of fixed and variable expense allocations for breeding and finishing subenterprises. We use recognized industry expense levels for each subenterprise to help accomplish expense separation.

Table 6.1. Fixed and variable expenses for hog enterprise, separated into breeding and finishing enterprises

Expense	Total	Breeding enterprise (60 female hogs)	Finishing enterprise (880 feeders)
Fixed expenses			
Management/labor	$4,500	$3,300	$1,200
Interest	5,745	4,218	1,527
Insurance	546	401	145
Taxes/utilities	480	352	128
Depreciation	7,200	5,925	1,275
Total	$18,471	$14,196	$4,275
Variable expenses			
Trucking/marketing	$3,150	$0	$3,150
Fuel/oil	1,560	415	1,145
Repair/maintenance	1,350	360	990
Supplies	520	140	380
Vet/medical	406	406	0
Interest	1,995	530	1,465
Total	$8,981	$1,851	$7,130

Note that we include an interest expense in both the fixed and variable expenses. When related to long-term debt incurred on fixed improvements, the interest expense is a fixed expense. The variable expense is that interest related to working capital debt.

Not included in either the fixed or variable expenses is the cost of feed. The cost of all rations should be calculated, and all feedstuffs delivered to each subenterprise must be measured separately. Surprisingly, many producers spend thousands of dollars to build bulk storage, central processing, and conveyor systems and forget to spend even $100 on feed measuring equipment—the most important of all functions.

Many of the expenses of Table 6.1 can be measured for separation. For instance, fixed interest is based on the financed value of each housing unit; depreciation can be calculated directly from the tax schedule; insurance and taxes can be measured; and management/labor time can be charted.

Analyzing the Hog Breeding Herd

Now that we've separated our expenses by subenterprise, we'll take a look first at the breeding herd and see how much it's costing this operator to maintain a female hog and wean a pig (Table 6.2). And as before, we must add the cost of feed to

Table 6.2. Breeding herd enterprise expenses

Expense	Total	Per female hog (60 head)	Per weaned pig (900 head)
Fixed expenses			
Management/labor	$3,300	$55.00	$3.67
Interest	4,218	70.30	4.69
Insurance	401	6.68	0.45
Taxes/utilities	352	5.87	0.39
Depreciation	5,925	98.75	6.58
Total	$14,196	$236.60	$15.78
Variable expenses			
Trucking/marketing	$0	$0	$0
Fuel/oil	415	6.92	0.46
Repair/maintenance	360	6.00	0.40
Supplies	140	2.33	0.16
Vet/medical	406	6.77	0.45
Interest (operating)	530	8.83	0.59
Total	$1,851	$30.85	$2.06

Table 6.3. Breeding expenses plus feed costs

	Total	Per female hog (60 head)	Per weaned pig (900 head)
Total fixed costs	$14,196	$236.60	$15.78
Total variable costs	1,851	30.85	2.06
Total feed cost	10,840	180.67	12.04
Total all costs	$26,887	$448.12	$29.88

the fixed and variable expenses (Table 6.3). It's extremely important to be able to identify the breeding herd and nursery feed separately from finishing feed.

Note that Tables 6.2 and 6.3 refer to weaned pigs. We could also separate the nursery from the breeding operation and produce the cost of a weaned and feeder pig separately. This should be done if the breeding herd is showing problems with cost.

Analyzing the Hog Finishing Herd

We have discovered a break-even cost of $29.87 to produce each 40-pound pig of the breeding herd (total breeding herd expenses [$26,887] divided by 900 pigs weaned and raised to 40 pounds). Knowing the cost of the breeding operation allows you to know exactly what it costs to finish the feeder pigs. Let's look at what happens when the 40-pound pigs are fin-

Table 6.4. Hog finishing enterprise expenses

Expense	Total	Per hog (880 head)	Per head day (120/head)
Fixed expenses			
Management/labor	$1,200	$1.36	$0.011
Interest	1,527	1.74	0.014
Insurance	145	.165	0.001
Taxes/utilities	128	.145	0.001
Depreciation	1,275	1.45	0.012
Total	$4,275	$4.86	$0.039
Variable expenses			
Trucking/marketing	$3,150	$3.58	$0.030
Fuel/oil	1,145	1.30	0.011
Repair/maintenance	990	1.12	0.009
Supplies	380	.43	0.004
Vet/medical	0	0	0
Interest (operating)	1,465	1.66	0.014
Total	$7,130	$8.09	$0.068

Table 6.5. Hog finishing expenses plus feed costs

	Total	Per hog	Per head day
Total fixed costs	$4,275	$4.86	$0.039
Total variable costs	7,130	8.09	0.068
Total feed cost	36,960	42.00	0.350
Total all costs	$48,365	$54.95	$0.457

ished to 240-pound hogs (Table 6.4). As with the breeding illustration (Table 6.3), we must add the cost of feed to the fixed
and variable expenses, as shown in Table 6.5. To identify the
cost of gain and conversion ratio, you must be able to segre-
gate the feed cost for each pen of hogs finished.

We have now identified and allocated all costs except for the
cost of death loss within the finishing barn. Of the 900 pigs
transferred, 880 were finished, producing a death loss of 20
hogs. These 20 hogs cost $29.87 as feeder pigs and will be
amortized among the 880 head sold at $0.679 per head (20
head×$29.87/880 head=$0.679 per head). This is a 2.2% death
loss, which is a bit high. As shown in Table 6.4, there were no
vet/medical costs in finishing. This producer might consider
that there should have been.

Having segregated all costs and identified the cost of each
function within the enterprise, we can now add them back together (Table 6.6).

Table 6.6. Total hog enterprise costs and cost per hog sold

	Total	Per hog sold
Total feeder pig costs	$26,887	$29.88
Total finishing costs	11,405	12.95
Finishing feed costs	36,960	42.00
Death loss costs	598	0.68
Total all costs	$75,850	$85.51

Using Your Break-even Cost

Table 6.6 shows that the total cost of sending a 240-pound hog to market is $85.51, or $35.63 per hundredweight. Very few operators can document a $35.63 per hundredweight breakeven, with most hog breakevens today around $40 or more. Most confinement systems have not generated the cost-saving efficiencies promised. Why? It's up to management to find out and make their systems more efficient. In the final analysis, management practices will determine whether an enterprise is profitable.

So the operator has a break-even cost of $35.63 per hundred weight. How does this benefit the operator if finished hogs are selling for $28.00 to $29.50 per hundredweight? This operation is still losing money—but not as much as it would if the breakeven were $42 per hundredweight.

Why do so few operators choose to exercise a price protection policy, and why are so many hogs sold at a loss? My theory is that it's terribly hard to get excited about protecting a commodity price if you don't know what it costs to produce the commodity. Let's face it, very few managers know their production costs. They spend so much time managing the physical labor that they forget about managing the financial side.

All that hard work for nothing. Given that the operator now knows the breakeven to be $35.63, is there anything that can be done with the current hog market? Yes! Table 6.7 illustrates some ways that one could use the futures market.

Having gotten past February 1995, one can lock in a profit for the balance of the year, if desirable. One could also purchase an APR $36 PUT, ensuring a loss of no more than $3.36 per hog; purchase an APR $36/40 PUT/CALL Window, offering a maximum profit of $8.64 per hog; purchase a JUN $40

Table 6.7. Live hogs (futures close for 12/5/94)

	Close	Profit (loss)/cwt
DEC 94	$30.92	$(4.71)
FEB 95	35.25	(0.38)
APR 95	36.67	1.04
JUN 95	42.07	6.44
JUL 95	42.15	6.52

PUT, at a profit of $6.96 per hog; or purchase a JUL $42 PUT at a profit of $9.29.

All illustrated prices of Table 6.7 are of course subject to a local cash basis adjustment. However, you can see that if one knows the cost of production, one can plan ahead to protect prices. It's the beginning of, and an absolute prerequisite to, an intelligent approach to marketing a perishable product in a mature marketplace.

Currently, many hog producers are concerned about whether they will be able to stay in business. And with good reason. Only those who are efficient, know their costs, and market wisely will survive into the future.

Dairy Enterprise Analysis

MOST DAIRY OPERATIONS in the United States today are still multienterprised operations and have the same problems identifying specific dairy-related costs as other types of farming. However, dairying has a slight advantage over some other enterprises in understanding and applying expenses because the industry has become accustomed to producing a wealth of herd- and production-specific operating facts. When you apply these facts correctly to gross income and expenses, you can discover a lot about how your business is operating.

Dairy Enterprise Expenses

Let's begin our analysis by reviewing the costs of the enterprise and dividing them into two common denominators: total cows (65 head) and hundredweight of milk sold (10,803 annual) (Table 7.1).

Table 7.1 does not include the cost of feed in either the fixed or variable expenses. As with some of the other enterprises already analyzed, the cost of all rations should be calculated as well as all feedstuffs delivered to each of the dairy herds—the lactating cows, dry cow/replacement herd, and growing calves. The total costs of feedstuffs and total operating costs from Table 7.1 generate a total cost of the enterprise as shown in Table 7.2.

The total cost of operations is $14.75 per hundredweight of

Table 7.1. Fixed and variable dairy enterprise expenses

Expense	Total	Per cow (65 cows)	Per cwt milk (10,803 cwt)
Fixed expenses			
Management/labor	$30,600	$470.77	$2.833
Insurance	2,340	36.00	0.217
Taxes/utilities	3,264	50.21	0.302
Interest	14,300	220.00	1.324
Depreciation	14,400	211.54	1.333
Total	$64,904	$998.52	$6.009
Variable expenses			
Fertilizer/chemicals	2,550	39.23	0.236
Trucking/marketing	11,550	177.69	1.069
Fuel/oil	2,964	45.60	0.274
Repair/maintenance	7,560	116.31	0.700
Supplies	2,600	40.00	0.241
Vet/medical	2,204	33.91	0.204
Interest (working capital)	2,900	44.61	0.268
Total	$32,328	$497.35	$2.992
Total operating costs	$97,232	$1,495.87	$9.001

Table 7.2. Dairy expenses plus feed costs

	Total	Per cow	Per cwt milk
Total operating costs	$97,232	$1,495.87	$9.001
Total feed costs	62,117	955.65	5.750
Total all costs	$159,349	$2,451.52	$14.751

milk sold. If milk is selling at an average of $12.85 per hundredweight, a few calves and cull cows must be sold to break even. Unfortunately, the figures of Table 7.2 are relatively close to the 1993 averages for most milk marketing areas in the United States, and when all hard and soft costs are considered, the industry, on average, appears to be operating at an approximate breakeven.

Obviously, this indicates that if one wishes to be successful in the dairy industry, one must operate at better than the average.

We noted the average dairy milk price of $12.85 per hundredweight because this was the average 1993 price calculated by the U.S. Department of Agriculture (USDA). Multiplying this price by 10,803 hundredweight of milk equals a total milk sale income of $138,819. The total costs of operation have already been calculated at $159,349, leaving the operation at a $20,530 income deficit.

If you were to calculate an average 20% herd replacement, producing the sale of 13 cull cows per year (13 cows×$500 per head=$6,500), and the sale of calves in excess of replacement needs (65 calves−15=50×$175 per calf=$8,750), you can generate an additional $15,250 of gross income. This additional income still leaves operations with a $5,280 loss for the year ($20,530 deficit−$15,250=$5,280). We noted that after all hard and soft costs are considered, the American dairy industry as a whole appears to be operating at no better than breakeven. Actually, the USDA figures would indicate a loss for the average producer in most milk markets, and that's what we're seeing here.

When faced with this situation, what do you do to solve the problem? Start by looking for the large (out-of-place) expenses. It helps to have per-cow benchmarks to compare your expenses with. In this case, those expenses that appear high are management/labor at $30,600 ($470.77 per cow), fixed interest at $14,300 ($220.00 per cow), and depreciation at $14,400 ($211.54 per cow). Without any other information available, these figures indicate high investment and debt per cow unit, with labor less than fully employed. When new facilities are being expanded or built, it's not unusual to overexpand. In fact, it's common to see confinement facilities built for 100 cows with a current population of only 70 to 80 cows. Note that all of the overstated expenses are fixed expenses and can be modified by changing the total number of cows. Under these circumstances, the underemployment of labor is common. If one good person is fully employed managing 90 cows, that same person will be underemployed with only 65 cows.

In total concept, we are talking about the low utilization of assets. As we've already stated, an operator must fully utilize assets to be profitable. So when you're faced with an unprofitable operation, determine first whether you're fully utilizing all assets.

Our hypothetical operator may have another problem as well. If we calculate the 10,803 hundredweight of milk sold on a per-cow basis, it equals a production of 16,620 annual pounds per cow. Once again, this is about average for the

United States as a whole. But the figures indicate that it's probably not a good idea to be just average. Again, the figures we think of as being high are the fixed expenses, and they can be modified by increasing cows and by increasing milk production per cow. Therefore, the operator might be able to make this operation profitable by increasing the number of cows, by increasing the milk produced from 16,620 to 18,000 or 20,000 pounds, or both.

Asset Utilization Problem

Increasing the number of cows

We'll make an example of asset utilization because in many farming enterprises, assets are not fully utilized, which has a direct bearing on profits.

Having explored two different problems that might be affecting profits adversely in our dairy operation, let's continue by pushing some figures to fully examine the effects of adding cows to the herd (Table 7.3). The operator currently milks 65 cows, is losing money, and the fixed expenses are too high per cow. Remember, the fixed expenses will change per cow as

Table 7.3. Changes in dairy expenses when cows are added to the herd

	Number of cows		
	65	75	85
Income			
Milk sales	$2,135.67	$2,135.67	$2,135.67
Cull/calves	234.61	234.61	234.61
Total	$2,370.28	$2,370.28	$2,370.28
Expenses			
Total fixed	$998.52	$865.40	$763.58
Variable expenses			
Fertilizer/chemical (pasture)	39.23	39.23	39.23
Trucking/marketing	177.69	177.69	177.69
Fuel/oil	45.60	45.60	45.60
Repair/maintenance	116.31	116.31	116.31
Supplies	40.00	40.00	40.00
Vet/medical	33.91	33.91	33.91
Interest (working capital)	44.61	44.61	44.61
Total	$497.35	$497.35	$497.35
Total feed costs	$955.65	$955.65	$955.65
Total all costs	$2,451.52	$2,318.40	$2,216.58
Profit (loss)/cow	$(81.24)	$51.88	$153.70

cows are added, but the variable expenses and feed costs will not.

It's hard to believe that a few cows can make the kind of difference shown in Table 7.3. But they do, especially when a producer has expensive facilities. Note that the fixed interest (interest paid on facilities, equipment, and cow debt) is $14,300 (Table 7.1). With a 9% fixed interest rate, the operation is currently amortizing $159,000 of debt. Also, depreciation is currently costing $14,400 per year (Table 7.1), and little of this is charged against the value of purchased cows because the operator raises the replacements. A 15-year capitalization factor would show that this operator is depreciating some $216,000 of facilities and equipment. That equals $3,323 per milk cow unit, plus the value of the raised cows, for a total of, say, $4,300 of capital structure per cow unit.

A lot of $12.85 per hundredweight milk is required to amortize the illustrated capital structure. That's why full utilization of assets is so important to any business that has high capital structure requirements, which includes a dairy operation. Table 7.4 reviews the profit (loss) per cow and total profit (loss) that can occur when the herd is increased.

Table 7.4. Profit (loss) with increased herd

	Number of cows		
	65	75	85
Profit (loss) per cow	$(71.24)	$51.88	$153.70
Total profit (loss)	$(4,631)	$3,891	$13,065

Increasing milk production

The example above showed that increasing the number of milk cows in the herd reduced fixed expenses per cow and increased profitability. This is fairly typical of an operation that is underutilizing assets. Also, we characterized this operator's milk production per cow as being low, although it's average for American producers. Let's examine how profits might increase with additional milk production. Table 7.5 shows figures for three production levels on a per hundredweight basis. Again, the fixed expenses will decrease per hundredweight as production goes up.

Table 7.5 shows that a few more pounds of milk per cow can make a significant difference.

So what are the differences between a cow that produces 16,000 pounds of milk and one that produces 20,000 pounds? The differences can be classified as genetic or related to feed ration, housing, medical care, and psychological comfort. One of these five differences, housing, can be characterized as being physical in nature and the others as being management oriented. The expenses of our hypothetical operation indicate very good housing facilities. As far as the management-oriented differences go, genetics have been established within the dairy industry that ensure 20,000 pounds of milk production. Controlling genetics is a matter of good breeding and culling practices. Medical care is not a problem in most operations; however, some could pay more attention to the psychological well-being of the animal by striving for a well-ordered, nonabrasive, calming atmosphere. Finally, a feed ration with a good trace mineral package and plenty of energy is extremely important to production.

Full utilization of assets can mean more than just having a barn full of cows. As noted, it can also mean getting the most out of each cow. Table 7.6 reviews the full effect of good milk

Table 7.5. Changes in dairy expenses when milk production is increased

	Annual milk per cow (lb)		
	16,620	18,000	20,000
Income			
Milk sales (per cwt)	$12.850	$12.850	$12.850
Cull/calves	1.412	1.412	1.412
Total (per cwt)	$14.262	$14.262	$14.262
Expenses			
Total fixed	$6.009	$5.547	$4.993
Variable expenses			
Fertilizer/chemicals (pasture)	0.236	0.236	0.236
Trucking/marketing	1.069	1.069	1.069
Fuel/oil	0.274	0.274	0.274
Repair/maintenance	0.700	0.700	0.700
Supplies	0.241	0.241	0.241
Vet/medical	0.204	0.204	0.204
Interest (working capital)	0.268	0.268	0.268
Total	$2.992	$2.992	$2.992
Total feed costs	$5.750	$5.750	$5.750
Total all costs (per cwt)	$14.751	$14.289	$13.735
Profit (loss)/cwt	$(0.489)	$(0.027)	$0.527

Table 7.6. Profit (loss) with increased milk production

	Lb of milk/cow		
	16,620	18,000	20,000
Profit (loss)/cwt	$(0.489)	$(0.027)	$0.527
Profit (loss)/cow	$(81.27)	$(4.86)	$105.40
Total profit (loss) for 65 cows	$(5,282)	$(315)	$6,851

production on profits. Full utilization can solve profit and cash flow problems.

Increasing both cows and milk production

In all farming enterprises, profitability is most often hindered when assets are not fully utilized. Our example dairy operation had a less-than-full dairy barn, which hinders profits in any operation and is devastating if the facilities are new and expensive. We looked at how increasing the herd would affect this operation and also discussed its milk production of 16,620 pounds per cow and the effect on profits of increasing per-cow production. Now let's see what happens if this operator were to increase both assets: the cows to 85 and milk production to 20,000 pounds (Table 7.7).

Our example illustrates a $30,154 turnaround in profitabil-

Table 7.7. Effect on dairy enterprise of increasing herd and milk production

	Annual milk (lb)/No. of cows	
	16,620/65	20,000/85
Income		
Milk sales	$138,819	$218,450
Cull/calves	15,250	19,942
Total	$154,069	$238,392
Expenses		
Total fixed	$64,904	$64,904
Variable expenses		
Fertilizer/chemicals (pasture)	2,550	4,012
Trucking/marketing	11,550	18,173
Fuel/oil	2,964	4,658
Repair/maintenance	7,560	11,900
Supplies	2,600	4,097
Vet/medical	2,204	3,468
Interest (working capital)	2,900	4,556
Total	$32,328	$50,864
Total feed costs	$62,117	$97,750
Total all dairy costs	$159,349	$213,518
Total profit (loss)	$(5,280)	$24,874

ity. And we haven't asked the operator to perform a miracle—just keep the barn full and exercise some good management practices. Maybe your barn is full, but if it isn't, think about it. Maybe you already get 20,000 pounds per cow, but most producers don't. If you're one of them, think about it and ask a nutritionist to balance your rations.

Cow/Calf Enterprise Analysis

NOW WE'LL EXPLORE the last of our hypothetical operator's five enterprises: the cow/calf enterprise. As with the other enterprises, the first task is to select common denominators by which expenses can be divided. In a beef cow operation the most obvious denominator is the number of cows bred and calves weaned. As analysis becomes more sophisticated, others might come to mind, such as pounds of beef weaned, weaned calf days, etc.

Cow/Calf Enterprise Expenses

Initially, we'll base our expense analysis on the number of cows bred (300 head) and calves weaned (265 head). Table 8.1 shows total fixed and variable expenses. In addition to the fixed and variable expenses, any feed costs other than pasture (winter feed cost) must be added. A simple method of calculating the cost is illustrated in Table 8.2. Once the feed costs are calculated, they can be added to the fixed and variable expenses, as in Table 8.3.

Although some cow/calf ranchers are single-enterprised, most have some grain farming/haying operations complicated enough to demand expense separation, and others background and/or finish cattle as well. Therefore, enterprising is as im-

Table 8.1. Fixed and variable cow/calf enterprise expenses

Expense	Total	Per cow (300 head)	Per calf (265 head)
Fixed expenses			
Management/labor	$17,100	$57.00	$64.53
Insurance	624	2.08	2.35
Rent (pasture)	19,075	63.58	71.98
Taxes/utilities	1,824	6.08	6.88
Depreciation	8,640	28.80	32.60
Total	$47,263	$157.54	$178.34
Variable expenses			
Fertilizer/chemicals (pasture)	$2,550	$8.50	$9.62
Trucking	1,050	3.50	3.96
Fuel/oil	936	3.12	3.53
Repair/maintenance	1,890	6.30	7.13
Supplies	325	1.08	1.23
Vet/medical	2,030	6.77	7.66
Interest	14,620	48.73	55.17
Total	$23,401	$78.00	$88.30
Total operating costs	$70,664	$235.54	$266.64

Table 8.2. Feed cost calculation

	No. of cows	Days on feed	Lb/day	Cost/lb	Total
Corn	300	60	8	$0.0402	$5,789
Silage	300	120	20	0.0100	7,200
Hay	300	150	40	0.0125	22,500
Total cost of feed					$35,489

Table 8.3. Cow/calf expenses plus feed costs

Expense	Total	Per cow	Per calf
Fixed	$47,263	$157.54	$178.34
Variable	23,401	78.00	88.30
Feed	35,489	118.30	133.92
Total	$106,153	$353.84	$400.56

portant to the beef cow rancher as it is to the balance of agriculture. As you can see, in either case the cow/calf operator should be careful to include the cost of winter feed in expenses.

Total Weaning Costs

As figured above, the total cost per calf weaned is $400.56. Is this all of the costs? No, not yet. A key indicator is the total depreciation cost of $8,640, or $28.80 per cow. As one *Profit*

Trend reader noted, "That doesn't sound like very much depreciation." It isn't, and the reason is that this operator raises replacement heifers rather than purchasing them. As such, depreciation is low, but it also means that the cost of weaning and raising replacements has not yet been fully considered. This is an additional cost of operating the cow herd, so let's examine the cost of cow replacements.

The first cost is the cost of weaning the calf that has been selected for breeding. One can use either the actual cost of $400.56 or that plus an opportunity cost equal to that which could have been received if the calf had been sold. Generally, the most conventional approach is to use actual cost.

To the first cost of weaning the calf one must add the cost of growing, maintaining, and breeding the calf during the next 12 to 24 months. For the sake of simplicity, let's say that these annual costs are as shown in Table 8.4 (for three-year-olds, multiply the values by two).

One could question whether we should also include a breeding charge in Table 8.4. This depends on where the cost of breeding is carried within the herd costs. In our example the breeding expenses are carried within the cost of the cow herd. Therefore, all costs have already been allocated on a per-cow-bred and per-calf-weaned basis. However, if the open heifers are totally separated from the cow herd, with separate bulls, etc., a breeding charge would be required. In our example 30 heifers are selected annually for breeding and are grown for two years before calving (three year olds), producing a cost of $780.56 per head, or $23,416.80 ($400.56+[$190×2]×30= $23,416.80).

This cost for any given set of bred heifers should be added to the income statement during the year when the bred heifers enter the cow herd, adding an additional cost to be amortized by the cows and weaned calves. With this we can determine a

Table 8.4. Annual costs for raising replacements

Expense	Amount
Pasture (annual)	$90.00
Winter feed	65.00
Vet/medical	15.00
Other	20.00
Total annual	$190.00

Table 8.5. Total costs for cow/calf herd

Expense	Total	Per cow	Per calf
Fixed	$47,263	$157.54	$178.34
Variable	23,401	78.00	88.30
Feed	35,489	118.30	133.92
Cow replacement	23,417	78.06	88.36
Total	$129,570	$431.90	$488.92

final cost for the cow/calf herd (Table 8.5). As always, if we neglect to consider all of the costs involved, we will have problems. Bad information negatively influences profit decisions all the way down the line.

Cow/Calf Income Statement

Now that we have achieved an accrual of all operating expenses for the cow/calf enterprise, let's find out whether the enterprise is profitable. This requires the building of an income (profit-and-loss) statement, and the only item missing is the income.

It's important to note that as in calculating the expense, we use the "accrual income" rather than "cash income" method. As such, the income will be calculated for the sale of all 265 calves even though only 235 are sold for cash after saving (holding back) 30 head of heifers for breeding. The 265 calves will be sold at 525 pounds, at $85 per hundredweight. The 30 cull cows are sold at $420 each. Table 8.6 shows the cow/calf

Table 8.6. Cow/calf income statement

	Total	Per cow	Per calf weaned
Income			
Calves (265 × 525 × $0.85)	$118,256	$394.19	$446.25
Cull cows (30 head × $420)	12,600	42.00	47.55
Total	$130,856	$436.19	$493.80
Expenses			
Fixed	$47,263	$157.54	$178.35
Variable	23,401	78.00	88.30
Winter feed	35,489	118.30	133.92
Replacement cost	23,417	78.06	88.36
Total	$129,570	$431.90	$488.92
Profit (loss)	$1,286	$4.29	$4.88

income statement. As illustrated, after all accrued income and expenses are considered, including the salary of the owner, the net profit is $1,286, or $4.88 per calf weaned. Therefore, selling calves at $85 per hundredweight is relatively close to breaking even. However, to calculate an exact break-even cost, follow the method shown in Table 8.7.

Obviously, in past years when 500- to 550-pound calves were bringing $95 to $105 per hundredweight, most cow/calf ranchers were making some money. However, at lower prices, many are just breaking even, and some are losing money.

But what happens if prices go down even farther, which could easily happen. Break-even costs have ranged from $45 to $139 per hundredweight of calves weaned. What happens will depend on each individual's cost of operating. For those who want to stay in business, now is the time to examine operating costs.

Weaning Numbers and Weights Make a Difference

In reviewing the cow/calf enterprise we have been using a herd of 300 cows and weaning 265 calves. This equals an 88.3% cow/calf weaning percentage, which is probably typical for the average producer. With these figures the average cost of weaning a calf is $441.37, which equals a break-even cost per hundredweight of $88.27 at a weaned weight of 500 pounds, $84.07 at a weight of 525 pounds, and $80.25 at 550 pounds. This variance of 50 pounds equals an $8.02 per hundredweight difference in break-even cost. That would get my attention. So what is the difference in an 85% and 95% weaned rate, and what is the value of weaning a 525-pound calf versus

Table 8.7. Calculation of break-even cost

	Total	Per cow	Per calf weaned
Operating costs	$129,570	$431.90	$488.92
Less cull sales	12,600	42.00	47.55
Net operating costs	$116,970	$389.90	$441.37
Cost of weaning			
500-lb calf			$88.27
525-lb calf			$84.07
550-lb calf			$80.25

a 475- or 550-pound calf? Table 8.8 reviews the total costs of operation, and Table 8.9 gives a qualitative and quantitative comparison.

As every cow/calf rancher knows, the comparison of Table 8.9 illustrates the need to maximize the percentage of calves weaned and the weaning weight of the calves if one is to be successful. Granted, maximizing both is not as important when calves are selling for $95 to $105 per hundredweight, but in today's atmosphere, it gets more important.

As illustrated, the cost moves upward approximately $10 per hundredweight when moving from a 95% to 85% weaning percentage. Also, the cost of weaning moves downward another $12 to $13 when moving from a 475- to 550-pound weaned weight. Therefore, this operation can change the break-even cost almost $22 per hundredweight depending on the number of calves weaned and the weaning weight.

This example can very quickly make a case for watching the cows closely at birthing time and feeding a few sacks of creep feed. These two items could easily make the difference for a lot of cow/calf producers during the next few years. We normally think of efficiency as being related to minimizing expenses or using assets wisely. But as illustrated, it can also be related to the operational aspects of the business. Spending money on creep feed can sometimes make an operation more efficient.

Weaning Cost Too High

With the analysis of our cow/calf enterprise in hand, one might say, "My weaning costs are too high." If so, what can we do to reduce the total cost, and consequently the cost per hundredweight?

Table 8.8. Total cost of operation

Expense	Total	Per calf (265 head)
Fixed	$47,263	$178.35
Variable	23,401	88.30
Feed	35,489	133.92
Net replacement	10,817	40.80
Total	$116,970	$441.37

Table 8.9. Weaned calf comparison

Weaned calves as percentage of total cows in herd (300 head)	95%	90%	88%	85%
No. weaned	285	270	265	255
Cost per calf	$410.42	$433.22	$441.37	$458.71
Cost/cwt weaned at				
475 lb	$86.40	$91.20	$92.92	$96.57
500 lb	82.08	86.64	88.27	91.74
525 lb	78.17	82.52	84.07	87.37
550 lb	74.62	78.77	80.25	83.40

We examined the benefits of weaning as many calves as possible and weaning large calves. A high weaned calf percentage depends somewhat on genetics and breeding practices but even more so on calving facilities and health management practices. The same is true for weaning large calves. Genetics and breeding influence the size of the calf, but herd health and nutritional management can dramatically affect the outcome. In the end, the most important ingredient of success is management, regardless of whether one is examining genetics, breeding, health, or nutrition.

Beyond herd management problems, one can also find fault with some of the expense figures we've been working with, which have nothing to do with the number or weight of calves weaned. It might not surprise anyone to know that the two expenses that come to mind first are related to the cost of cows and pasture. Table 8.1 shows a pasture cost of $63.58 per cow. However, not all of the pasture is rented. The operator owns 1,600 acres and rents approximately 2,000 acres at the rate of $9.50 per acre. The pasture is rated at a capacity of 12 acres per cow/calf unit. Therefore, the rented pasture is really costing $114 per cow/calf unit. It's only when averaged over the total cow herd that the net pasture rent equals $63.58 per head. In recent years many ranchers have had to pay $110 to $120 per unit for rented pasture. Obviously, this is the result of competition for grass within the industry; however, it has not been very economical for the cow/calf rancher and has increased the cost of weaning a calf.

The other major factor of cost increase during the past several years is the cost of replacement cows and heifers. Before a modest decrease during the latter part of 1995, replacements were priced at $900 to $1,100 per head. Paying these prices for replacements has increased the interest expense dramatically. Interest cost (assuming a 9% interest rate) is $14,620, or $48.73 per cow unit (Table 8.1), with the total debt per cow equal to $541.44.

Granted, the total debt may consist of a combination of cow and land purchase debt. When one is required to amortize debt on both cows and land, the cost of weaning goes up. The point is that within this present cycle, the cost of cows and land (whether purchased or rented) has increased to a virtual breakeven for most ranchers. While going through the bottom of the cycle, many costs will have to be reduced.

9

Analysis Summary

IN OUR LOOK at enterprise (profit center) analysis, our example has been a five-enterprise farming operation that included the cropping of corn, soybeans, and alfalfa hay; a cattle feeding enterprise; a hog farrow-to-finish operation; a dairy operation; and a cow/calf enterprise. This drive to evaluate all of the enterprises began with an analysis of financial ratios and other indicators and the recognition that these ratios and indicators can give management the wrong message—in our case, of nonprofitability. In fact, the balance sheet ratios were rapidly falling apart.

Most lenders today are still balance sheet lenders, and a balance sheet that's falling apart will probably dictate future borrowing problems; therefore, it's important not only to recognize that the balance sheet is declining but to understand *why* and *what can be done about it.*

As we pointed out, financial ratios and indicators are important tools in recognizing the level of profitability, or lack thereof, and the resulting increase in or deterioration of balance sheet values. However, you can't begin to understand fully the *why* (what is wrong with operations) and the *what can be done about it* (to create profitability) without subjecting the entire business to a detailed review such as we've illustrated in our enterprise (profit center) analysis. We hope that our exercise has given you a new appreciation for the value of analyzing profitability.

Cropping Summary

The cropping enterprise analysis illustrated that 120 bushels per acre of corn was being harvested at a break-even cost of $2.43 per bushel, 35 bushels of soybeans at $6.86 per bushel, and 4 tons of alfalfa hay at $31.58 per ton. Granted, our analysis already included full management/labor charges that equaled $23.80 per corn acre ($0.198 per bushel), but that's a legitimate expense whether paid to the grocery store or paid in the form of a salary check. What must be recognized is that very little corn raised at this cost had been sold for a profit over the past 10 years. If this producer were in the federal set-aside program, he/she would have received an annual payment averaging approximately $0.30 per bushel, reducing the net cost to $2.13 per bushel.

In the end it can be argued that most producers profited only to the extent of part of the annual set-aside payments on corn raised during the past 10 years, which is borne out in our illustration. This could be one reason why most balance sheets are sitting still—or falling apart.

In our analysis of cropping operations, we reviewed the bushels or tons per acre of production for each crop and the break-even cost of producing each crop. When we review the break-even cost, it's not hard to deduce whether we can sell the product for a price in excess of cost, but it leaves us wanting for just how this affects the balance sheet.

Remember, we originally introduced enterprise analysis because "the balance sheet was falling apart," and we wanted to find out how and why. To boil this down to something that directly affects the balance sheet, all we need is an income statement. Table 9.1 illustrates some input factors, which then are used to develop some income statement figures.

A small profit of $9,118 has been produced. We could also add any U.S. Department of Agriculture payments to the profit. To do so we would have to recalculate the remaining acres of production after a 5% or 10% set-aside. Was a profit produced because of what most of us would consider to be the two major crops? No—it was because of the alfalfa acres. It's typical for alfalfa to produce a higher profit potential.

The only atypical factor in our example is for acres to be

Table 9.1. Cropping enterprise input factors and income statement figures

	Corn	Soybeans	Alfalfa	Total
Input factors				
Total acres	334	333	333	1000
Bu/ton per acre	120	35	4	. . .
Total cost/acre	$291.96	$240.27	$126.34	. . .
Sale price/unit	$2.30	$6.00	$50.00	. . .
Income statement				
Total income	$92,184	$69,930	$66,600	$228,714
Total cost	97,515	80,010	42,071	219,596
Profit (loss)	$(5,331)	$(10,080)	$24,529	$9,118
Per acre	$(15.96)	$(30.27)	$73.66	$9.12
Per bu/ton	$(0.133)	$(0.865)	$18.42	

planted one-third each to corn, soybeans, and alfalfa hay. A more typical planting schedule would devote approximately 50% (500 acres) to corn, 30% (300 acres) to soybeans, and 20% or less (200 acres) to alfalfa hay. Table 9.2 shows what happens to profitability if we use the same input factors and introduce this more typical cropping schedule. You can see that the per-acre and per-unit figures stay the same. Gross income increases $12,286, and net profit decreases $11,447.

We've analyzed cropping operations and discovered that after an entire year of work and investing $219,596, it all comes down to a pretax profit of just over $9,000. At best, this is not a pretty picture, and it's one of the reasons why the balance sheet is "heading south." At this point, anyone with an ounce of common sense would ask, "Is there anything I can do about this?" Yes, there are options to increase profits.

A good example is a producer who farms mostly irrigated ground. Several years ago this producer decided that the normal corn/soybeans rotation with a little alfalfa hay for the cows wasn't doing the job. He decided to get into the alfalfa hay

Table 9.2. Effect on income statement of more typical cropping schedule

	Corn	Soybeans	Alfalfa	Total
Total income	$138,000	$63,000	$40,000	$241,000
Total cost	145,980	72,081	25,268	243,329
Profit (loss)	$(7,980)	$(9,081)	$14,732	$(2,329)
Per acre	$(15.96)	$(30.27)	$73.66	. . .
Per unit	$(0.133)	$(0.865)	$18.42	. . .

business as a cash crop and today devotes 50% of his ground to alfalfa. This producer sells hay for $80 to $100 per ton and has tripled his net income. Table 9.3 shows a comparision using 50% hay acres.

The net income has *tripled*, increasing $19,732, and the total equals 316% of the original $9,118 profit. Even more striking, the return on investment for corn is 5% and for alfalfa is 58%. And this is using a sale price of $50 per ton for hay and nonirrigated farming. Table 9.4 shows what would happen if we used $85 per ton and irrigated the ground. The striking results may seem unbelievable—but they're true.

Crop Breakevens Affect Cattle Feeding

When analyzing a diversified farming operation, one can expect to see much of the harvested crop consumed by livestock feeding and multiplier herds. Indeed, one of the positive theories of diversified operations is that one can more easily raise livestock with inexpensively produced grains and forages.

Some producers have come to think of internally produced feeds as a "cheap" feedstuff supply that allows them to grow livestock when others cannot afford to and even hold animals past the point of finish—commonly called "out-feeding a poor market." To test this theory, we'll feed our example livestock to get a picture of just how it affects them. You'll remember our example of 660 head of cattle fed annually from approximately 750 to 1,250 pounds. To begin, we must establish the rations to be used for this weight bracket. Table 9.5 lists the

Table 9.3. Effect of using 50 percent hay acres

	Corn	Alfalfa	Total	Total
Input factors				
Total acres	500	500	1000	For corn/soyb/
Bu/ton per acre	120	4	...	alf rotation
Total cost/acre	$291.96	$126.34	...	(1/3-1/3-1/3)
Sale price/unit	$2.30	$50.00	...	...
Income statement				
Total income	$138,000	100,000	$238,000	$228,714
Total cost	145,980	63,170	209,150	219,596
Profit (loss)	$(7,980)	$36,830	$28,850	$9,118

Table 9.4. Effect of $85 per ton hay sale price and irrigated ground

	Corn	Alfalfa	Total
Irrigated input factors			
Total acres	500	500	1000
Bu/ton per acre	155	6.5	...
Total cost/acre	$341.96	$196.34	...
	(Incr $50)	(Incr $70)	
Sale price/unit	$2.30	$85.00	...
Income statement			
Total income	$178,250	276,250	$454,500
Total irrigated cost	170,980	98,170	269,150
Profit (loss)	$7,270	$178,080	$185,350
Return on investment	4.3%	181.0%	69.0%

pounds of each ingredient per ton of ration, the dry matter conversion, total ingredients fed to each steer at an 8-to-1 conversion, the break-even cost of ingredients produced by the farm, and a cost per each ton of ration.

For ease of comparison, in Table 9.6 silage is costed at 10 times the price of corn. The first column illustrates all ingredients at the harvested cost and the second at current purchase price; the third assumes corn raised at a price of $1.80 per bushel. On just a small 300-head feedlot, the difference between $2.43 and $2.16 for corn is $8,190, or $12.41 per head, with a savings of $35.10 per head using $1.80 for corn. This is sometimes the only profit in cattle feeding. You can see that the cost of cropping does have a dramatic effect on the profitability of livestock enterprises.

Feed-to-Gain Conversion and Breakeven

Above we examined feeding differently priced feedstuffs to cattle and how the price of the ingredients affected the feed

Table 9.5. Cattle ration breakdown

Ingredients	Lb/ton	Dry matter content	Total lb fed[a]	Cost/unit	Cost/ton
Corn (80%)	1,600	1,360	3,040	$2.43	$69.43
Silage	150	75	285	24.30	1.82
Alfalfa	150	135	285	31.59	2.37
Concentrate	100	90	190	145.00	7.25
Total	2,000	1,660	3,800	...	$80.87

[a]Based on an 8-to-1 conversion.

Table 9.6. Comparative feed ration cost, based on price variation

	Harvested cost	Cost/head	Current purchase price	Cost/head	Corn at $1.80/bu	Cost/head
Corn	$(2.43)	$131.91	$(2.16)	$117.26	$(1.80)	$97.71
Silage	(24.30)	3.46	(21.60)	3.08	(18.00)	2.56
Alfalfa	(31.59)	4.50	(50.00)	7.12	(31.59)	4.50
Concentrate	(145.00)	13.77	(145.00)	13.77	(145.00)	13.77
Total		$153.64		$141.23		$118.54
×660 head		$101,402		$93,212		$78,236
Total savings				$8,190		$23,166
Savings/head				$12.41		$35.10
Cost/cwt of gain		$32.35		$29.73		$24.96

cost of gain on a per-hundredweight basis. Discussing the importance of the cost of producing a crop, and the resulting effects this has on the feeding of livestock, brings us to the point of examining what happens to the feed once it has been fed: the conversion of feed into red meat.

If you are knowledgeable about feeding cattle, you may have reviewed the conversion ratio of 8 to 1 (on an as-fed basis) and the resulting $32.35 cost per hundredweight of gain figure (using $2.43 per bushel corn) with a bit of skepticism. As well you should. An 8-to-1 "as-fed basis" conversion ratio is extremely efficient and can be met by some cattle feeders on some occasions, but not by many very often. The ration illustrated has a 0.83 dry matter basis, which equals a dry matter basis conversion ratio of 6.6 to 1. Many variables enter into the ability to convert feed efficiently, including weather conditions, housing, and genetics as well as the quality of feed ingredients and management of all other factors.

As illustrated, these variables should be considered carefully in regard to how they affect the bottom-line cost of producing a finished product. Table 9.7 compares 8-to-1, 9-to-1, and 10-to-1 conversions on an "as-fed basis." Our producer is growing corn at a cost of $2.43 per bushel, silage at $24.30 per ton, and alfalfa hay at $31.59 per ton; you will note that all ingredients are being used at the cost of production. Therefore, the ration is costing $80.87 per ton as fed to the cattle.

Including an original purchase cost of $550 per head, or $70.97 per hundredweight, the effects of the comparative conversion ratios can be summarized as in Table 9.8.

Table 9.7. Comparison of three different conversions on an as-fed basis

	Ration cost		Feed-to-gain conversion		
	Per unit	Per ton	8 to 1	9 to 1	10 to 1
Corn	$2.43	$69.43	$131.91	$148.40	$164.89
Silage	24.30	1.82	3.46	3.89	4.32
Alfalfa	31.59	2.37	4.50	5.06	5.63
Concentrate	145.00	7.25	13.77	15.49	17.21
Total feed cost/head fed			$153.64	$172.84	$192.05
Total feed (lb as-fed)			3,800	4,275	4,750
(dry matter basis)			3,154	3,548	3,942
Dry matter basis conversion ratio			6.6/1	7.5/1	8.3/1
Feed cost/cwt of gain			$32.34	$36.39	$40.43
All other costs/cwt[a]			18.84	18.84	18.84
Total all costs/cwt			$51.18	$55.23	$59.27

[a]All other costs are calculated at a 3.3 average daily gain, on feed 150 days, with total non-feed costs of $0.472 per head day—a total of $89.49 per head.

Table 9.8. Summary of effects of comparative conversion ratios

	Conversion ratio		
	8 to 1	9 to 1	10 to 1
Original purchase price	$550.00	$550.00	$550.00
Total cost of gain	243.10	262.34	281.53
Total all costs	$793.10	$812.34	$831.53
Break-even sales price (cwt)	$ 63.45	$64.99	$66.52
Increase (cwt)	...	$1.54	$3.07

Crop Breakevens Affect the Hog Enterprise

Crop breakevens have the same effect on the hog farrow-to-finish enterprise. A farrow-to-finish operation does not normally consume as wide a variety of grains and forage feedstuffs as cattle, although the breeding stock rations contain some ground alfalfa meal. However, all hog rations are heavy on corn (or a similar grain), and today some producers process their own soybeans. Table 9.9 shows a typical set of female hog, feeder pig, and finishing rations used in a farrow-to-finish operation, with raised corn used as the energy input ingredient. For our purposes soybean meal will be included commercially. As you can see from the table, using the producer's break-even cost of $2.43 per bushel for corn, the current approximate purchase cost of $2.16, and an assumed excellent raised cost of $1.80 per bushel produces a dramatic difference

Table 9.9. Hog farrow-to-finish rations (ingredient list and pounds per ton of ration)

	Female hog rations		Feeder pig	Grower	Finisher
	Gestation	Lactation	Feeder pig	Grower	Finisher
Corn	1,715	1,570	1,204	1,565	1,660
Soybean meal	205	360	450	380	285
Concentrate	70	65	300	52	52
Trace mineral	10	5	46	3	3
Total	2,000	2,000	2,000	2,000	2,000
Percent corn	86%	79%	60%	78%	83%
Corn cost in ration					
@ $2.43/bu	$73.74	$67.51	$51.77	$67.29	$71.38
@ $2.16/bu	$66.88	$61.23	$46.96	$61.03	$64.74
@ $1.80/bu	$54.88	$50.24	$38.53	$50.08	$53.12

Table 9.10. Total ration and corn consumed annually[a]

	Female hog rations		Feeder pig	Grower	Finisher
	Gestation	Lactation	Feeder pig	Grower	Finisher
Lb fed/day	6	10	1.5	3.9	6.5
Days on feed	317	48	30	60	60
Total feed	1,902	480	45	234	390
Corn fed/head	1,636	379	27	183	324
No. of head fed	1	1	15	14.67	14.67
Total corn used	1,636	379	405	2,685	4,753

[a]Assumes two female hog cycles, 15 weaned pigs, 14.67 finished.

in the corn cost per ton of ration. With this cost differential, we can examine the effect on operations for each female hog unit and the finishing of her offspring, as shown in Tables 9.10 and 9.11. The savings illustrated are for each and every female hog unit in a farrow-to-finish operation. As you can see, the cost of producing corn has a dramatic effect on the cost of a hog enterprise.

Breakevens Make or Break an Operation

Few hog producers go to the field to plant corn remembering that if corn can be produced for $2.16 rather than $2.43 per bushel the savings differential equals $47.53 per female hog unit, and if produced at a nominal $1.80 per bushel the savings equal $110.91 per female hog unit. In case the savings differ-

Table 9.11. Total corn cost and savings differential

Per bushel	$2.43	$2.16	$1.80
Total corn cost	$427.77	$380.24	$316.86
Savings differential	$0	$47.53	$110.91

Table 9.12. Savings differential for corn produced at $2.16 and $1.80 per bushel

	$2.16/bu	$1.80/bu
Per female hog unit	$47.53	$110.91
100 Units	$4,753	$11,091
200 Units	$9,506	$22,182
500 Units	$23,765	$55,455

ential hasn't sunk in yet, Table 9.12 shows how this adds up to a lot of money. Producers can multiply the savings by the number of female hogs in their operation.

Owners of hog operations have switched vets for less than the savings shown in Table 9.12, changed feed companies for less, fired herd managers for less, and spent $1,500 per female hog on new buildings for less. Put another way, a $47.53 savings per female hog will pay 100% of the interest cost on most 100-, 200-, and 500-female hog operations.

In this case the problem is not the fault of the vet, the feed company, the herd manager, or the buildings. The fault is farming enterprises that do not complement each other. During the past three decades, certain cropping expenses have gotten out of line enough to totally negate the advantages of a diversified farming operation. When the price of producing corn during a normal production year starts to exceed $2.10 per bushel, the producer should consider selling the combine and custom hiring the harvest, selling the "big power" and renting a field tractor for 45 days, and paying less for rent per acre and buying the corn.

Poor cropping breakevens, and crops that consequently cost too much to produce, affect the breakevens in the livestock to which the crops are fed. Thus, the inefficiency and lack of productivity of one enterprise are transferred to another, until eventually the entire operation is down the drain.

The relationship of cropping and livestock breakevens can

be examined by illustrating the changing cost of corn on the per-hundredweight cost of producing pork (Table 9.13; the illustration is based on producing 240-pound hogs). As shown, the $0.27 difference between a $2.43 and $2.16 bushel of corn produces a $1.35 per hundredweight savings in total feed cost. If you divide the $0.27 into the $1.35 savings, the answer is five bushels of corn. These five bushels times 2.4 hundredweight of pork produced per hog equals a total of 12 bushels, including maintenance of the breeding herd.

Breakevens Affect Costs of Dairy and Cow/Calf Enterprises

On average, the entire dairy industry today is in an approximate break-even position. As such, a few effective producers are making good money, many are barely getting by, and many are losing money. Milk pricing supports provided by the U.S. Department of Agriculture also establish a stable but static

Table 9.13. Relationship between costs of corn and producing pork

	Corn cost/bu		
	$2.43	$2.16	$1.80
Per female hog unit	$427.77	$380.24	$316.86
Per finished hog	$29.16	$25.92	$21.60
Per cwt of pork	$12.15	$10.80	$9.00
Savings/cwt	$0	$1.35	$3.15

Table 9.14. Relationship between costs of corn and dairying

	Corn cost/bu		
	$2.43	$2.16	$1.80
Cost/head day	$0.955	$0.849	$0.707
Annual cost (305 days)	$291.27	$258.94	$215.63
Cost/cwt of milk	$1.46	$1.29	$1.08
Savings			
Per head day		$0.106	$0.248
Per cwt of milk		$0.170	$0.380
Annual savings for			
100 dairy cows		$3,233	$7,564
150 dairy cows		$4,849	$11,346
500 dairy cows		$16,165	$37,820

pricing structure that varies little from one market to another. Therefore, daily control of expenses is very important to the profit-minded dairy operator.

The largest daily cost of a dairy cow is feed expense, and the most important and expensive ingredient is for energy—the cost of corn. Therefore, the cost of cropping is an extremely important consideration in a dairy farm. As illustrated in Table 9.14, the cost of corn in dairy rations can make all the difference in bottom-line profits. The example assumes a dairy cow on a lactation ration for 305 days per year, eating 22 pounds of corn per day, and producing 20,000 pounds of milk annually.

You can readily determine the extreme importance of feed cost in a dairy operation. However, one normally would not assess the same importance to a beef cow/calf operation. After all, the beef cow spends most of the year on grass with a small amount of supplement at best. Only during the winter months within the last 90 to 120 days before calving are cows fed an energy ration. However, just because the feeding period is only 90 to 120 days doesn't mean it doesn't affect the cost of a weaned calf. To illustrate, let's assume a cow eating 10 pounds of corn per day for 90 days, and an 85% weaning rate and 500-pound weight. Table 9.15 shows that it does make a measurable difference: almost $1.00 per hundredweight of weaned calf for $2.16 versus $2.43 per bushel of corn, and $2.33 per hundredweight for $1.80 per bushel of corn. The cattle industry is presently close to the bottom of a cow/calf cycle, and with reduced calf prices many producers will want to reduce operating costs. How many ranchers have taken a look at the cost of cropping? Most producers don't regularly calculate the cost of raised feeds. If they don't, high cropping costs represent an insidious, silent thief of profits. Quite often, it's too late when finally noticed.

Table 9.15. Corn cost comparative

	Corn cost/bu		
	$2.43	$2.16	$1.80
Cost/cow	$39.06	$34.74	$28.89
Cost/cwt weaned calf	$8.98	$7.99	$6.65
Savings/cwt weaned calf		$0.99	$2.33

The Interdependence of Enterprises

In certain parts of the United States and Canada, a major percentage of all energies is devoted to pure, cash grain farming operations. And to such operations, a discussion of the dependency of livestock operations on cost-effective crop production has little meaning. However, a major cross section of North American agriculture is still diversified to include both cropping and livestock operations. Therefore, we have spent considerable time tying the success of livestock enterprises to the cost of cropping—the cost of producing homegrown feed ingredients for the livestock. As we've shown, the livestock are indeed very dependent on cropping costs.

Does the interdependence run in both directions? Does the success of the cropping enterprise also depend on how the livestock enterprises are managed? Indeed, yes! The success of a diversified farming operation depends on all enterprises contributing to the profit potential of the sum total.

To illustrate, it's of little value for an operator to reduce the cost of producing corn from $2.43 to $2.16 per bushel if the corn fed to finishing cattle or hogs is not included in a ration that achieves a high rate of feed-to-gain conversion. In our cattle feeding example, a steer fed to 1,200 pounds consumes approximately 53 bushels of corn (along with other ingredients) at a conversion ratio of 8 to 1, and the savings of $0.27 per bushel equals $14.31 per steer fed. However, this entire savings disappears if the conversion ratio increases a mere 0.8 pound per day, from an 8-to-1 to an 8.8-to-1 conversion—evidence that it is important to feed well-balanced rations.

The negative economic effects felt from a poor conversion ratio are the same for the production of pork, poultry, milk, eggs, or lamb chops. Likewise, effective crop production costs are of little value if the livestock products produced are not purchased and/or sold effectively. In conclusion, all enterprises within a farming operation are interdependent for overall profitability.

A balanced and well-managed diversified farming operation including crop and livestock operations is economically beneficial to the owner/operator. However, many of the benefits have been lost over time through either ineffective management or unbalanced operations. Can a producer be profitable with ineffective management? No, not to any great extent. The degrees of poor management and nonprofitability are proportional. An unbalanced operation can be defined as one that was once purposefully designed for balanced diversification and through either internal or external influences no longer exists as such.

Over the past 20 years many diversified operators have been converted to cash grain operations. Many were influenced by a lender and/or outside advisor, almost all of whom lacked the benefit of sound operating knowledge. However, in the end, it is the owner/operator's responsibility to understand the operation well enough to defend its existence.

The biggest problem in agriculture today is the net loss of producers. According to U.S. Department of Agriculture statistics, 34% of all producers left the business in the 1980s, and there isn't any reason to believe that less than one-third will leave in the 1990s. Furthermore, these lost producers are not being replaced by the next generation. Some experts argue that farming is simply too expensive to get into. This is largely true and is caused by arbitrarily inflated land and equipment costs, which in turn are caused by "safety net" programs of the federal government, including set-aside acres, the Conservation Reserve Program (CRP), loans, etc.

However, one can also argue that the lack of profitability is as great a factor—if not greater—in the stability and future of agriculture. Profitability can be determined only by analyzing past performance and transferring that knowledge to future operations. We've had our "wake-up call." Agriculture needs each producer to become more concerned with operating analysis, full utilization of assets, operating efficiency, and— *profitability*.

⟳ Index